Artificial Intelligence and the Future of Humanity

Other Books in the Current Controversies Series

America's Role in a Changing World
Big Tech and Democracy
The Business of College Sports: Who Earns What
The Capitol Riot: Fragile Democracy
Climate Change and Biodiversity
Cyberterrorism
Domestic Extremism
Media Trustworthiness
The Politics and Science of COVID-19
Reparations for Black Americans
Space: Tourism, Competition, Militarization

Artificial Intelligence and the Future of Humanity

Lisa Idzikowski, Book Editor

Published in 2023 by Greenhaven Publishing, LLC
29 E. 21st Street
New York, NY 10010

Copyright © 2023 by Greenhaven Publishing, LLC

First Edition

All rights reserved. No part of this book may be reproduced in any form without permission in writing from the publisher, except by a reviewer.

Articles in Greenhaven Publishing anthologies are often edited for length to meet page requirements. In addition, original titles of these works are changed to clearly present the main thesis and to explicitly indicate the author's opinion. Every effort is made to ensure that Greenhaven Publishing accurately reflects the original intent of the authors. Every effort has been made to trace the owners of the copyrighted material.

Cover image: metamorworks/Shutterstock.com

Library of Congress Cataloging-in-Publication Data

Names: Idzikowski, Lisa, editor.
Title: Artificial intelligence and the future of humanity / Lisa Idzikowski, Book Editor.
Description: First edition. | New York, NY : Greenhaven Publishing, 2023. | Series: Current controversies | Includes bibliographical references and index. | Audience: Ages 15+ | Audience: Grades 10-12 | Summary: "Anthology of diverse viewpoints exploring developments in artificial intelligence, predictions for the technology's future, and the impact it will have on life as we know it"— Provided by publisher.
Identifiers: LCCN 2021060473 (print) | LCCN 2021060474 (ebook) | ISBN 9781534508866 (library binding) | ISBN 9781534508859 (paperback) | ISBN 9781534508873 (ebook)
Subjects: LCSH: Artificial intelligence—Juvenile literature.
Classification: LCC Q335.4 .A235 2023 (print) | LCC Q335.4 (ebook) | DDC 006.3—dc23/eng20220215
LC record available at https://lccn.loc.gov/2021060473
LC ebook record available at https://lccn.loc.gov/2021060474

Manufactured in the United States of America

Website: http://greenhavenpublishing.com

Contents

Foreword	**10**
Introduction	**13**

Chapter 1: Does Artificial Intelligence Pose a Threat to Humanity?

Overview: The Potential of Artificial Intelligence Is Uncertain   **17**

> *Peter Stratton and Michael Milford*
>
> Stratton and Milford generally agree that artificial intelligence will result in positive outcomes. The authors present a balanced outlook on the many possibilities of improvements on today's uses of AI and project what may be in store for the future of AI.

Yes: Artificial Intelligence Poses a Threat to Humanity

Will the Singularity Be Humanity's Last Invention?   **23**

> *Martin Kaste*
>
> Some in the tech community believe that computers will someday be smarter than humans. So, what would happen then? It is speculated that these smart computers will become so superior to humans that they will end the human race.

Technology Favors Rich Nations Over Poor Nations   **29**

> *Cristian Alonso, Siddharth Kothari, and Sidra Rehman*
>
> Developed nations already have a distinct advantage over poorer, developing nations when it comes to technology. This occurs because of a variety of factors. The authors suggest that developing nations must act quickly to lessen the negative effects.

Artificial Intelligence Poses a Threat to Privacy   **33**

> *Sarah Ovaska*
>
> Consumers are aware of the potential privacy risks when it comes to increased use of artificial intelligence, and they want their privacy to be protected. This viewpoint analyzes ways that companies can achieve this end. The author argues that Europe is further advanced in this area.

No: Artificial Intelligence Does Not Pose a Threat to Humanity

Use Artificial Intelligence to Increase Food
Production and Curb Poverty   40

> *Lisa Rabasca Roepe*
>
> Poverty and hunger are major problems in parts of the world. Artificial intelligence can be used to alleviate both problems. Major research in the US is determining how to tackle food shortages by refining two major crops: sorghum and soybeans.

AI Is Positive for Education Now and in the Future   44

> *William Goddard*
>
> AI is proving to be a positive influence on education, potentially even more so in the future. A wide range of educational benefits is explored by the author, who also finds a small number of potential negative aspects.

People Encounter Helpful AI Every Day   52

> *Sasha Reeves*
>
> How many times a day does the average person use AI? Some people probably don't realize how imbedded AI is in their lives and how much they rely on it. Eight common applications of AI are explored in this viewpoint.

Chapter 2: Will Artificial Intelligence Change What It Means to Be Human?

Overview: Advances in Technology Could Already Put
Evolution into Hyperdrive   58

> *David Trippett*
>
> Over time, evolution has slowly changed human beings into what they are today. But what could happen if technology interfered in that process? Would people accept a change in their bodies or minds?

Yes: Technology Will Change Human Beings

There Are Differences Between Human Intelligence
and Artificial Intelligence   62

> *J. E. Korteling, G. C. van de Boer-Visschedijk, R. A. M. Blankendaal, R. C. Boonekamp, and A. R. Eikelboom*
>
> Should we pursue the development of AI "partners" with human-level intelligence or should we focus on supplementing human limitations? In order to answer these questions, humans working

with AI systems in the workplace or in policy making have to develop an adequate mental model of the underlying psychological mechanisms of AI, according to this excerpted viewpoint.

Robots Will Be Human-Like   76

Progressive Automations Inc.

Companies around the world are developing human-like robots, and these machines keep getting better and better. Robots are used now in a variety of situations, and this trend will increase the usage into the future.

No: Technology Will Mimic Human Beings

Human Intelligence Is Superior to AI in Specific Ways   85

Ben Dickson

It is unquestionable that AI can process data much better than humans. But in certain ways, human intelligence far outperforms computers at present. AI and human intelligence are so different that it's not always a good idea to compare the two.

AI Will Be Human-Like, but Better   90

Ana Santos Rutschman

The late Stephen Hawking was conflicted about artificial intelligence. He warned against superhuman AI and its potential to threaten humanity but saw the value of AI to prevent, fight, and lessen the burden of disease. Until AI gets away from us, it has the power to improve countless lives.

Chapter 3: Will Artificial Intelligence Change the World of Work?

Overview: AI's Influence on the Labor Market Will Require Changes to Education   95

Elizabeth Mann Levesque

Education will be the key to unlocking future workforce stability in the face of changes brought about by AI. This education must start in the elementary years and continue throughout an individual's life.

Yes: AI Will Change the Working World in Significant Ways

A Significant Loss of Jobs Will Occur in the Future   103

Calum McClelland

There is no doubt that with the continuing advancement of AI, many of today's jobs will disappear. This revolution in the world of work has occurred at other times throughout history.

Work Will Be Transformed by Artificial Intelligence   112
Dennis Spaeth

The labor force will be transformed by artificial intelligence. Many experts believe that the largest change will be the loss of jobs that require continuous, repetitive skills. Unfortunately, this will affect many workers.

No: AI Will Not Change Everything in the Workplace

Automation Doesn't Necessarily Mean Massive Job Loss   116
Carlos Bonilla

Automation anxiety is nothing new. There has been a cycle of automation anxiety, the fear of no jobs left for human workers, several times in American history. Only time will tell, but it is a good bet that the worry of job losses is overblown as in the past.

People and Machines Will Work Together in the Future   121
Justin Lokitz

A very positive aspect of AI and the world of work is that people and machines will increasingly work together. Tasks for humans will be easier as machines assist people in many areas of work.

Chapter 4: Should Artificial Intelligence Be Controlled for the Future Good of Society?

Overview: How Could Artificial Intelligence Be Regulated?   126
Oren Etzioni

Everyone knows that AI is not something that will go away. In fact, governments and leaders know that AI has the potential to provide immeasurable power. The question is how to keep it safe.

Yes: AI Should Be Controlled or Regulated to Reduce Negative Effects

Perfection of AI Must Not Overtake Human Empathy   132
Arshin Adib-Moghaddam

To allow AI and its perfection to overtake human empathy would be a serious mistake. The author emphatically declares that AI must be controlled or regulated so that this would not happen because dire consequences could result.

Cybersecurity Is a Top Concern in Artificial Intelligence   **135**

Josephine Wolff

In a time of rapidly evolving threats, cybersecurity is a vital concern in the field of AI. Governments must realize that national security is not the only area in which AI must be regulated and controlled. In fact, AI systems themselves must be secured.

Control AI to Combat Fake News and Disinformation   **143**

Darrell M. West

Fake news and distorted reality, especially on social media sites, is eroding democracy. Experiments have shown that most people cannot spot fake news when it is presented online. According to this excerpted viewpoint, the solution is to control and regulate AI to prevent fake news in the first place.

No: Regulation of AI Should Capitalize on Increasing Its Positive Effects

The Case for Improving National Security   **154**

Alex Ciarniello

The case for using AI to improve national security is straightforward. Current commercial use is helping to model systems for national security use, and this will and must be enhanced for the future.

Continue Improving Autonomous Vehicles to Save Lives   **158**

Teena Maddox

Autonomous vehicles will continue to improve in safety and efficiency. Experts insist that with the improvement in autonomous vehicles and the infrastructure to support them, traffic fatalities will measurably decrease.

Global Citizens Need an AI-Infused Curriculum   **162**

UNESCO

UNESCO seeks to address worldwide peace initiatives through its programs that focus on equity in education, working to ensure that AI does not widen technological divides.

Organizations to Contact	**168**
Bibliography	**172**
Index	**174**

Foreword

"Controversy" is a word that has an undeniably unpleasant connotation. It carries a definite negative charge. Controversy can spoil family gatherings, spread a chill around classroom and campus discussion, inflame public discourse, open raw civic wounds, and lead to the ouster of public officials. We often feel that controversy is almost akin to bad manners, a rude and shocking eruption of that which must not be spoken or thought of in polite, tightly guarded society. To avoid controversy, to quell controversy, is often seen as a public good, a victory for etiquette, perhaps even a moral or ethical imperative.

Yet the studious, deliberate avoidance of controversy is also a whitewashing, a denial, a death threat to democracy. It is a false sterilizing and sanitizing and superficial ordering of the messy, ragged, chaotic, at times ugly processes by which a healthy democracy identifies and confronts challenges, engages in passionate debate about appropriate approaches and solutions, and arrives at something like a consensus and a broadly accepted and supported way forward. Controversy is the megaphone, the speaker's corner, the public square through which the citizenry finds and uses its voice. Controversy is the life's blood of our democracy and absolutely essential to the vibrant health of our society.

Our present age is certainly no stranger to controversy. We are consumed by fierce debates about technology, privacy, political correctness, poverty, violence, crime and policing, guns, immigration, civil and human rights, terrorism, militarism, environmental protection, and gender and racial equality. Loudly competing voices are raised every day, shouting opposing opinions, putting forth competing agendas, and summoning starkly different visions of a utopian or dystopian future. Often these voices attempt to shout the others down; there is precious little listening and considering among the cacophonous din. Yet listening and considering, too, are essential to the health of a democracy. If controversy is democracy's lusty lifeblood, respectful listening and careful thought are its higher faculties, its brain, its conscience.

Current Controversies does not shy away from or attempt to hush the loudly competing voices. It seeks to provide readers with as wide and representative as possible a range of articulate voices on any given controversy of the day, separates each one out to allow it to be heard clearly and fairly, and encourages careful listening to each of these well-crafted, thoughtfully expressed opinions, supplied by some of today's leading academics, thinkers, analysts, politicians, policy makers, economists, activists, change agents, and advocates. Only after listening to a wide range of opinions on an issue, evaluating the strengths and weaknesses of each argument, assessing how well the facts and available evidence mesh with the stated opinions and conclusions, and thoughtfully and critically examining one's own beliefs and conscience can the reader begin to arrive at his or her own conclusions and articulate his or her own stance on the spotlighted controversy.

This process is facilitated and supported in each Current Controversies volume by an introduction and chapter overviews that provide readers with the essential context they need to begin engaging with the spotlighted controversies, with the debates surrounding them, and with their own perhaps shifting or nascent opinions on them. Chapters are organized around several key questions that are answered with diverse opinions representing all points on the political spectrum. In its content, organization, and methodology, readers are encouraged to determine the authors' point of view and purpose, interrogate and analyze the various arguments and their rhetoric and structure, evaluate the arguments' strengths and weaknesses, test their claims against available facts and evidence, judge the validity of the reasoning, and bring into clearer, sharper focus the reader's own beliefs and conclusions and how they may differ from or align with those in the collection or those of classmates.

Research has shown that reading comprehension skills improve dramatically when students are provided with compelling, intriguing, and relevant "discussable" texts. The subject matter of these collections could not be more compelling, intriguing, or urgently relevant to today's students and the world they are poised to inherit. The anthologized articles also provide the basis for stimulating, lively, and passionate classroom debates. Students who are compelled to anticipate objections to their own argument and identify the flaws in those of an opponent read more carefully, think more critically, and steep themselves in relevant context, facts, and information more thoroughly. In short, using discussable text of the kind provided by every single volume in the Current Controversies series encourages close reading, facilitates reading comprehension, fosters research, strengthens critical thinking, and greatly enlivens and energizes classroom discussion and participation. The entire learning process is deepened, extended, and strengthened.

If we are to foster a knowledgeable, responsible, active, and engaged citizenry, we must provide readers with the intellectual, interpretive, and critical-thinking tools and experience necessary to make sense of the world around them and of the all-important debates and arguments that inform it. We must encourage them not to run away from or attempt to quell controversy but to embrace it in a responsible, conscientious, and thoughtful way, to sharpen and strengthen their own informed opinions by listening to and critically analyzing those of others. This series encourages respectful engagement with and analysis of current controversies and competing opinions and fosters a resulting increase in the strength and rigor of one's own opinions and stances. As such, it helps readers assume their rightful place in the public square and provides them with the skills necessary to uphold their awesome responsibility—guaranteeing the continued and future health of a vital, vibrant, and free democracy.

Introduction

> "AI is a tool that will be used by humans for all sorts of purposes, including in the pursuit of power. There will be abuses of power that involve AI, just as there will be advances in science and humanitarian efforts that also involve AI."
>
> *Danah Boyd, founder of the Data & Society Research Institute*[1]

Pondering the future of humanity feels a bit like accepting a challenge to count the number of sand grains in the Sahara Desert or the stars in the universe. A daunting task at best. But it is interesting to imagine what humanity may become as well as the challenges facing society within the world of artificial intelligence (AI). Merriam-Webster defines artificial intelligence as "a branch of computer science dealing with the simulation of intelligent behavior in computers" or "the capability of a machine to imitate intelligent human behavior."

Some may dismiss AI as science fiction, "too far out there," or only the domain of big tech companies. Many people may be surprised to understand that they use and depend on AI from the time they get up until the time they go to bed: First thing in the morning, tap into social media to see what is happening, later on check Google maps to see if there is a traffic backup, before leaving school or work do some online banking, and after eating dinner crash on the couch while surfing Netflix for a good movie.

According to the International Science Survey done by Pew Research in 2019–2020, AI is generally viewed as a good thing for society: 47 percent of Americans said AI was good versus 44 percent that disagreed. Global views on this topic vary widely, with 72 percent of the citizens of Singapore viewing AI in a positive light and only 18 percent of Japanese indicating it was negative.

It is worth considering the question whether AI could possibly be harmful to human society. Various scientific experts and computer gurus have weighed in on this subject. SpaceX CEO Elon Musk has more than once cited the potential dangers he feels are inherent in AI development. At a tech conference in 2018, he said, "I am very close to the cutting edge in AI and it scares the hell out of me." And the late theoretical physicist professor Stephen Hawking, undeniably one of the world's greatest thinkers, told the BBC in 2014 that "the development of full artificial intelligence could spell the end of the human race."

Of course, not all computer experts or scientific minds believe that AI poses an existential threat to human society. Many point to its positive contributions to life in the present and potential for the future. Anyone who has flown on an airplane or has watched any number of spaceflights probably marvels at the capabilities of technology. Who would have thought that going into space would ever be possible for ordinary people, instead of just for highly trained astronauts? And who doesn't know of an elderly individual in need of personal care? Japan has been a leader in robot manufacturing and since 2015 has placed robots in nursing homes around the country to help take care of elderly residents.

Is it possible that AI will change society or what it means to be human? Experts have various opinions on the topic. Some fear that the world of work will never be the same when AI is widely implemented. Proponents of AI in the workplace predict that when repetitive, low-level jobs are taken over by robots or smart machines, higher skilled jobs will need to be filled by humans. Opponents argue that too many workers will be displaced by AI and that society will need to find meaningful employment for those workers. In any case, Americans have differing views of the situation: 31percent of women who responded to Pew's International Science Survey believed that AI automation in the workplace was a negative compared to 52 percent of men, who saw it as a positive thing.

What about transhumanism? Britannica defines the topic as a "movement devoted to promoting the research and development of robust human-enhancement technologies." These would include computer or digital augmentation of human sensory capabilities, cognitive capacity, human health, and even the ability to extend human life spans. Of course, there are some people presently living as a result of highly technical medical procedures and/or medically adapted hardware. On an extreme level though, pro-transhumanists believe that in the future humans might become immortal. Interestingly, in a survey by scientists from the University of Texas and reported in the *Journal of Aging Studies* in June 2021, only 33 percent of those individuals surveyed were interested in the ability to take a pill that would make them immortal.

Obviously, artificial intelligence has already impacted human society and will continue to do so in the future. Questions abound whether the technology and its research should be put under close watch and regulation. There are many arguments supporting both points of view. Some people suggest that to overly regulate the industry would deny society of almost unimaginable benefits, or that the negative consequences are overstated. On the other hand, knowledgeable individuals believe that to leave the industry to its own devices might ultimately doom our species.

The current debate surrounding this fascinating topic can be investigated and comprehended by reading and contemplating the diverse viewpoints contained in *Current Controversies: Artificial Intelligence and the Future of Humanity*.

Notes

1. "Artificial Intelligence and the Future of Humans," by Janna Anderson and Lee Rainie, Pew Research Center, December 10, 2018.

Chapter 1

Does Artificial Intelligence Pose a Threat to Humanity?

Overview: The Potential of Artificial Intelligence Is Uncertain

Peter Stratton and Michael Milford

Peter Stratton is a research fellow at the University of Queensland, Australia, and Michael Milford is a professor at the Queensland University of Technology.

Artificial intelligence (AI) promises to revolutionise our lives, drive our cars, diagnose our health problems, and lead us into a new future where thinking machines do things that we're yet to imagine.

Or does it? Not everyone agrees.

Even billionaire entrepreneur Elon Musk, who admits he has access to some of the most cutting-edge AI, said recently that without some regulation "AI is a fundamental risk to the existence of human civilization."

So what is the future of AI? Michael Milford and Peter Stratton are both heavily involved in AI research and they have different views on how it will impact on our lives in the future.

How Widespread Is Artificial Intelligence Today?

Michael: Answering this question depends on what you consider to be "artificial intelligence."

Basic machine learning algorithms underpin many technologies that we interact with in our everyday lives—voice recognition, face recognition—but are application-specific and can only do one very specific defined task (and not always well).

More capable AI—what we might consider as being somewhat smart—is only now becoming widespread in areas such as online

---

"The Future of Artificial Intelligence: Two Experts Disagree," by Peter Stratton and Michael Milford, The Conversation, July 17, 2017. https://theconversation.com/the-future-of-artificial-intelligence-two-experts-disagree-79904. Licensed under CC BY-ND-4.0 International.

retail and marketing, smartphones, assistive car systems and service robots such as robotic vacuum cleaners.

Peter: The most obvious and useful examples of current AI are the speech recognition on your phone, and search engines such as Google. There is also IBM's Watson, which in 2011 beat human champion players at the US TV game show *Jeopardy* and is now being trialled in business and healthcare.

Most recently, Google's DeepMind AI called AlphaGo beat the world champion Go player, surprising a lot of people—especially since Go is an extremely complex game, way surpassing chess.

What Major Advances in AI Will We See over the Next 10 Years?

Peter: Many auto manufacturers and research institutions are competing to create practical driverless cars for general road use. While currently these cars can drive themselves for much of the time, many challenges remain in dealing with bad weather (heavy rain, fog and snow) and random real-world events such as roadworks, accidents and other blockages.

These incidents often require some degree of human judgement, common sense and even calculated risk to successfully navigate through. We are still a long way from fully autonomous vehicles that don't need a licensed driver ready to take control in an instant.

The same can be said for all the AI that we will see over the coming 10-20 years, such as online virtual personal assistants, accountants, legal and financial advisers, doctors and even physical shop-bots, museum guides, cleaners and security guards.

They will be advanced tools that are very useful in specific situations, but they will never fully replace people because they will have little common sense (probably none, in fact).

Michael: We will definitely see a range of steady, incremental improvements in everyday AI. Online product recommendations will get better, your phone or car will understand your voice increasingly well and your vacuum cleaner robot won't get stuck as often.

It's likely that we'll see some major advances beyond today's technology in some but not all of the following areas: self-driving cars, healthcare, utilities (electricity, water, and so on) management, legal, and service areas such as cleaning robots.

I disagree on self-driving cars—there's no real reason why there won't be fully autonomous controlled ride-sharing fleets in the affluent centres of cities, and this is indeed the strategy of companies such as NuTonomy, working in Singapore and Boston.

What Approaches Will Lead to the Biggest Improvements in AI?

Michael: Major advances will come from two sources.

First, there is a long runway of steady incremental improvements left in many areas of conventional AI—large, complex neural networks and algorithms. These systems will continue to improve steadily as more training data becomes available and as scientists perfect them.

The second area will likely be biological inspiration. Scientists are only just starting to tap into the knowledge about how brain networks work, and it's likely they will copy or adapt what we know about animal and human brains to make current deep learning networks far more capable.

Peter: Old-fashioned AI, which was based on pure logic and computer programs that tried to get machines to behave intelligently, basically failed to do anything that humans are good at and computers are not (speech and image recognition, playing complex strategic games, for example).

What's quite clear now is that our best-performing AI is based on how we think the brain works.

But our current brain-based AI (called Deep Artificial Neural Networks) is still light years away from emulating an actual brain. Enhanced AI capabilities in the future will come from developing better theories of how the brain works.

The fundamental science needed to cultivate these theories will probably come from publicly funded research institutions, which will then be spun off into commercial start-up companies, and then quickly acquired by interested large corporations if they look like they might be successful.

How Will Artificial Intelligence Affect Society and Jobs?

Peter: Most jobs won't be under threat for a long time, probably several generations. Real people are needed to actually make any significant decisions because AI currently has no common sense.

Instead of replacing jobs, our overall quality of life will go up. For example, right now few people can afford a personal assistant, or a full-time life coach. In the near future, we'll all have (a virtual) one!

Our virtual doctor will be working for us daily, monitoring our health and making exercise and lifestyle suggestions.

Our houses and workplaces might be cleaner, but we will still need people to clean the spots the robots miss. We'll also need people to deploy, retrieve and maintain all the robots.

Our goods will be cheaper due to reduced transport costs, but we'll still need human drivers to cover all the situations the self-drivers can't.

All this doesn't even mention the whole new entertainment technologies and industries that will spring up to capture our increased disposable income and to cash-in on our improved quality of life.

So yes, jobs will change, but there will still be plenty of them.

Michael: It's likely that a significant fraction of jobs will be under threat over the coming decade. It's important to note that this won't necessarily be divided by blue-collar versus white-collar, but rather by which occupations are easily automatable.

It's unlikely that an effective plumber robot will be built in the near future, but aspects of the so far undisrupted construction industry may change radically.

Some people say machines will never have the emotional capabilities of humans. Whether that is true or not, many jobs will be under threat with even the most rudimentary levels of emotional understanding and interaction.

Don't think about the complex, nuanced interaction you had with your psychologist; instead think about the one with that disinterested, uncaring part-time hospitality worker. The bar for disruption is not as high as many think.

That leaves the question of what happens then. There are two scenarios—the first being that, like in the past, new types of jobs are generated by the technological revolution.

The other is that humanity gradually transitions into a Utopian society where scientific, artistic and sporting pursuits are pursued at leisure. The short to medium-term reality is probably somewhere in between.

Will Skynet/the Machines Take Over and Enslave Humanity?

Michael: It's unlikely in the near future but possible. The real danger is the unpredictability. Skynet-like killer cyborgs as featured in the Terminator film series are unlikely because that development cycle takes a while, and we have multiple opportunities to stop development.

But AI could destroy or damage humanity in other unpredictable ways. For example, when big companies like Google Deepmind start entering into healthcare, it's likely that they will improve patient outcomes through a combination of big data and intelligent systems.

One of the temptations or pressures will be to deploy these extremely complex systems before we completely understand every possible ramification. Imagine the pressure if there is good evidence it will save thousands of lives per year.

As we well know, we have a long history of negative unintended consequences with new technology that we didn't fully understand.

In a far-fetched but not impossible healthcare scenario, deploying AI may lead to catastrophic outcomes—a world-wide

AI network deciding in ways invisible to us human observers to kill us all off to optimise some misguided performance goal.

The challenge is that with newly developing technologies, there is an illusion of 100% control, which doesn't really exist.

Peter: All our current AI, and any that we can possibly create in the foreseeable future, are just tools—developed for specific jobs and totally useless outside of the exact duties they were designed for. They don't have thoughts or feelings. These AIs are just as likely to try to take over the world as your Xbox or your toaster.

One day, I believe, we will build machines that rival us in intelligence, and these machines will have their own thoughts and possibly learn in an unconstrained way. This sounds scary. But humans are dangerous for exactly the reasons that the machines won't be.

Humans evolved in a constant struggle for life and death, which made us innately competitive and potentially treacherous. When we build the machines, we can instead build them with any underlying motivation that we would like.

For example, we could build an intelligent machine whose only desire is to dismantle itself. Or, we could build in a hidden remote-controlled off switch that is completely separate from any of the machine's own circuits, and an auto-shutdown reflex if the machine somehow ever notices it.

All these safeguards will be trivial to implement. So there is simply no way that we could accidentally build a machine that then tries to wipe out the human race.

Of course, because humans themselves are dangerous, someone could build a machine that doesn't have these safeguards and use it for nefarious purposes. But we have that same problem now with nuclear weapons.

In the future, just as now, we have to hope that we are simply smart enough to use our technology wisely.

Will the Singularity Be Humanity's Last Invention?

Martin Kaste

Martin Kaste is a correspondent for NPR, covering issues relating to law enforcement, privacy, and major world events.

It's been called "the rapture of the nerds." For some computer experts, the Singularity is the moment when an artificial intelligence learns how to improve itself in an exponential "intelligence explosion." They say it's a bigger threat to puny humans than global warming or nuclear war—and they're trying to figure out how to stop it.

ROBERT SIEGEL, host: From NPR News, this is ALL THINGS CONSIDERED. I'm Robert Siegel.

How do you think life, as we know it, will end? Nuclear war? Climate change? How about an out-of-control computer?

(Soundbite of movie, "2001: A Space Odyssey")

Mr. DOUGLAS RAIN (Actor): (as HAL 9000) I know I've made some very poor decisions recently, but I can give you my complete assurance that my work will be back to normal.

SIEGEL: That, of course, is HAL 9000 from Stanley Kubrick's science fiction masterpiece "2001: A Space Odyssey."

Well, in 2011, some people think we're getting closer to inventing an artificial intelligence that could figure out how to make itself smarter. If so, they say, it might be the last thing humans ever invent.

NPR's Martin Kaste has the story.

© 2011 National Public Radio, Inc. NPR news report titled "The Singularity: Humanity's Last Invention?" was originally broadcast on NPR's *All Things Considered* on January 11, 2011, and is used with the permission of NPR. Any unauthorized duplication is strictly prohibited.

MARTIN KASTE: There's an apartment in downtown Berkeley where they're trying to save the world.

(Soundbite of knocking)

KASTE: Hello. It's four apartments, actually, which have been rented by something called the Singularity Institute for Artificial Intelligence.

Mr. KEEFE ROEDERSHEIMER (Software Engineer, Singularity Institute for Artificial Intelligence): Hi, how is it going?

KASTE: Good. Thank you.

Mr. ROEDERSHEIMER: Can I offer you guys some tea?

KASTE: Keefe Roedersheimer is one of the institute's research fellows. Over cups of green tea, he explains that he's a software engineer who's done work for NASA, and that his idea of a good time is teaching a computer how to play poker like a human.

But right now, at the institute, he's trying to predict the rate of advancement of artificial intelligence or A.I.

Mr. ROEDERSHEIMER: So it's about knowing when this could happen.

KASTE: By this, he's talking about the invention of a computer that's not only smart but also capable of improving itself.

Mr. ROEDERSHEIMER: Is able to look at its own source code and say, ah, if I change this, I'm going to get smarter. And then by getting smarter, it sees new insights into how to get smarter. And then by having those insights into how to get smarter, it modifies its source code and gets smarter and gets some insights. And that creates an extraordinarily intelligent thing.

KASTE: They call this the A.I. singularity. Because the intelligence could grow so fast, human minds might not be able to keep up. And therein lies the danger.

You've already seen this movie.

(Soundbite of movie, "Terminator 2: Judgment Day")

Mr. ARNOLD SCHWARZENEGGER (Actor): (as The Terminator) Skynet begins to learn at a geometric rate. It becomes self-aware at 2:14 a.m. Eastern time, August 29th. In a panic, they try to pull the plug.

Ms. LINDA HAMILTON (Actress): (as Sarah Connor) Skynet fights back.

Mr. SCHWARZENEGGER: (as The Terminator) Yes.

KASTE: They kind of hate it at the institute when you quote the "Terminator," but Roedersheimer says, at least, those movies gave people a sense of what could happen.

Mr. ROEDERSHEIMER: That's an A.I. that could get out of control. But if you really think about it, it's much worse than that.

KASTE: Much worse than "Terminator"?

Mr. ROEDERSHEIMER: Much, much worse.

KASTE: How could it possibly—that's a moonscape with people hiding under burnt out buildings and being shot by laser. I mean, what could be worse than that?

Mr. ROEDERSHEIMER: All the people are dead.

KASTE: In other words, forget the heroic human resistance. There'd be no time to organize one. Somebody presses enter, and we're done.

The singularity idea has floated around the edges of computer science since the 1960s, but these days, it's the subject of Silicon Valley philanthropy.

At a fund-raising party in San Francisco, the co-founder of PayPal, Peter Thiel, explains why he supports the Singularity Institute.

Mr. PETER THIEL (Co-Founder, PayPal): People are not worried about what supersmart computers will do to change the world, because we don't see those every day. And so I suspect that there are a lot of these issues that are being underestimated.

KASTE: Also at the party is Eliezer Yudkowsky, the 31-year-old who co-founded the institute. He's here to mingle with potential new donors. As far as he's concerned, preparing for the singularity takes primacy over other charitable causes.

Mr. ELIEZER YUDKOWSKY (Research Fellow and Director, Singularity Institute for Artificial Intelligence): If you want to maximize your expected utility, you try to save the world and the future of intergalactic civilization instead of donating your money to the society for curing rare diseases and cute puppies.

KASTE: Yudkowsky doesn't have formal training in computer science, but his writings have a following among some who do. He says he's not predicting that the future super A.I. will necessarily hate humans. It's more likely, he says, that it'll be indifferent to us—but that's not much better.

Mr. YUDKOWSKY: While it may not hate you, you're made of atoms that it can use for something else. So it's probably not a good thing to build that particular kind of A.I.

KASTE: What he and the institute are trying to do, he says, is start the process of figuring out how to build what he calls friendly A.I. before somebody inevitably builds the unfriendly variety.

But that day still seems a long way off when you look at the current state of A.I.

Good morning. Hello?

Unidentified Female: Are you looking for Eric?

KASTE: A computerized receptionist guards the office of Microsoft distinguished scientist Eric Horvitz.

Unidentified Female: Eric is working on something now. I think he won't mind too much, though, if you interrupt him. Would you like to go in?

KASTE: Horvitz is past president of the Association for the Advancement of Artificial Intelligence. He's working on systems that can greet visitors, do basic medical diagnoses and even read human body language.

Mr. ERIC HORVITZ: One whole direction we're going in is to bring together machine vision, machine learning, conversational abilities to explore what we call integrative A.I. And this is one path to brighter intelligences some day.

KASTE: But Horvitz doubts that one of these virtual receptionists could ever lead to something that takes over the world. He says that's like expecting a kite to evolve into a 747 on its own.

So does that mean he thinks the singularity is ridiculous?

Mr. HORVITZ: Well, no. I think there's been a mix of views, and I have to say that I have mixed feelings myself.

KASTE: In part because of ideas like the singularity, Horvitz and other A.I. scientists have been doing more to look at some of the ethical issues that might arise over the next few years with narrow A.I. systems.

They've also been asking themselves some more futuristic questions. For instance, how would you go about designing an emergency off switch for a computer that can redesign itself?

Mr. HORVITZ: I do think that the stakes are high enough where even if there was a low, small chance of some of these kinds of scenarios, that it's worth investing time and effort to be proactive.

KASTE: Still, many see the Singularity Institute and like-minded organizations as fringe. One computer scientist let slip the word cultish; others mock the singularity as the rapture of the nerds.

At the institute, they shrug this off. As far as they're concerned, it's just a matter of being rational about the future—relentlessly rational.

Jasen Murray, for instance, says he has no illusions about the institute's ability to succeed at its mission.

Mr. JASEN MURRAY (Program Manager, Singularity Institute for Artificial Intelligence): We have between 30 and 60 years to figure out this—to solve this ridiculously hard problem that we probably have a low chance of solving correctly and—ah, this is just really bad.

KASTE: But they're willing to try. The institute is looking to move out of its apartments in Berkeley and buy a big old Victorian house. That way, its researchers can have a more permanent home for whatever time humanity has left.

Martin Kaste, NPR News.

Technology Favors Rich Nations Over Poor Nations

Cristian Alonso, Siddharth Kothari, and Sidra Rehman

Cristian Alonso, Siddharth Kothari, and Sidra Rehman are economists at the International Monetary Fund (IMF).

New technologies like artificial intelligence, machine learning, robotics, big data, and networks are expected to revolutionize production processes, but they could also have a major impact on developing economies. The opportunities and potential sources of growth that, for example, the United States and China enjoyed during their early stages of economic development are remarkably different from what Cambodia and Tanzania are facing in today's world.

Our recent staff research finds that new technology risks widening the gap between rich and poor countries by shifting more investment to advanced economies where automation is already established. This could in turn have negative consequences for jobs in developing countries by threatening to replace rather than complement their growing labor force, which has traditionally provided an advantage to less developed economies. To prevent this growing divergence, policymakers in developing economies will need to take actions to raise productivity and improve skills among workers.

Results from a Model

Our model looks at two countries (one advanced, the other developing) that both produce goods using three factors of production: labor, capital, and robots. We interpret "robots"

"How Artificial Intelligence Could Widen the Gap Between Rich and Poor Nations," by Cristian Alonso, Siddharth Kothari, and Sidra Rehman, IMF Blog, December 2, 2020. Reprinted by permission from the International Monetary Fund.

broadly, to encompass the whole range of new technologies mentioned above. Our main assumption is that robots substitute for workers. The "artificial intelligence revolution" in our framework is an increase in the productivity of robots.

We find that divergence between developing and advanced economies can occur along three distinct channels: share-in production, investment flows, and terms-of-trade.

Share-in-production: Advanced economies have higher wages because total factor productivity is higher. These higher wages induce firms in advanced economies to use robots more intensively to begin with, especially when robots easily substitute for workers. Then, when robot productivity rises, the advanced economy will benefit more in the long run. This divergence grows larger, the more robots substitute for workers.

Investment flows: The increase in productivity of robots fuels strong demand to invest in robots and traditional capital (which is assumed to be complementary to robots and labor). This demand is larger in advanced economies due to robots being used more intensively there (the "share-in-production" channel discussed above). As a result, investment gets diverted from developing countries to finance this capital and robot accumulation in advanced economies, thus resulting in a transitional decline in GDP in the developing country.

Terms-of-trade: A developing economy will likely specialize in sectors that rely more on unskilled labor, which it has more of compared to an advanced economy. Assuming robots replace unskilled labor but complement skilled workers, a permanent decline in the terms of trade in the developing region may emerge after the robot revolution. This is because robots will disproportionately displace unskilled workers, reducing their relative wages and lowering the price of the good that uses unskilled labor more intensively. The drop in relative price of its main output, in turn, acts as a further negative shock, reducing the incentive to invest and potentially leading to a fall not just in relative but in absolute GDP.

Robots and Wages

Our results critically depend on whether robots indeed substitute for workers. While it may be too early to predict the extent of this substitution in the future, we find suggestive evidence that this is the case. In particular, we find that higher wages coincide with significantly higher use of robots, consistent with the idea that firms substitute away from workers and towards robots in response to higher labor costs.

Implications

Improvements in the productivity of robots drive divergence between advanced and developing countries if robots substitute easily for workers. In addition, those improvements will tend to increase incomes but also increase income inequality, at least during the transition and possibly in the long run for some groups of workers, in both advanced and developing economies.

There is no silver bullet for averting divergence. Given the fast pace of the robot revolution, developing countries need to invest in raising aggregate productivity and skill levels more urgently than ever before, so that their labor force is complemented rather than substituted by robots. Of course, this is easier said than done. In our model, increases in total factor productivity—which account for the many institutional and other fundamental differences between developing and advanced countries not captured by labor and capital inputs—are especially beneficial as they incentivize more robots and physical capital accumulation. Such improvements are always beneficial, but the gains are stronger in the context of the artificial intelligence revolution.

Our findings also underscore the importance of human capital accumulation to prevent divergence and point to potentially different growth dynamics among developing economies with different skill levels. The landscape is likely going to be much more challenging for developing countries which have hoped for high dividends from a much-anticipated demographic transition. The growing youth population in developing countries was hailed by policymakers as possibly a big chance to benefit from a transition of jobs from China as a result of its graduating middle-income status. Our findings show that robots may steal these jobs. Policymakers should act to mitigate those risks. Especially in the face of these new technologically-driven pressures, a drastic shift to rapidly improve productivity gains and invest in education and skills development will capitalize on the much-anticipated demographic transition.

Artificial Intelligence Poses a Threat to Privacy

Sarah Ovaska

Sarah Ovaska is a freelance writer based in the United States.

Artificial intelligence (AI) has the potential to solve many routine business challenges—from quickly spotting a few questionable charges in thousands of invoices to predicting consumers' needs and wants.

But there may be a flipside to these advances. Privacy concerns are cropping up as companies feed more and more consumer and vendor data into advanced, AI-fuelled algorithms to create new bits of sensitive information, unbeknownst to affected consumers and employees.

This means that AI may create personal data. When it does, "it's data that has not been provided with [an individual's] consent or even with knowledge," said Chantal Bernier, assistant and interim privacy commissioner in the Office of the Privacy Commissioner of Canada from 2008 until 2014 who now consults in the privacy and cybersecurity practice of global law firm Dentons.

AI is an umbrella term used to describe advanced technologies such as machine learning and predictive analytics that essentially shift decisions once solely made by humans to computers.

While AI is still in its early stages—we may have robotic vacuums but nothing like the futuristic cartoon character Rosey, the robot maid from *The Jetsons*—industries are using the technology to expand revenue streams and reduce workforce costs by linking disparate bits of information.

Few corporate executives are focused on the privacy risks associated with the use of AI. Discussions in boardrooms and in C-suites are "more focused on the possibilities and the benefits

---

"Data Privacy Risks to Consider When Using AI," by Sarah Ovaska, originally appeared in *FM Magazine*, Feb/Mar 2020. Copyright 2020, Association of International Certified Professional Accountants. Reprinted by permission.

of AI than the potential risks," said Imran Ahmad, a Toronto-based lawyer with Blake, Cassels and Graydon who specialises in technology and cybersecurity issues.

Customers Want Assurances

Consumers are paying more attention to their private information and becoming increasingly uneasy about how data about their interests, locations, credit histories, and more is used by entities they interact with.

Seventy-one per cent of respondents surveyed by global professional services firm Genpact in 2018 said they don't want companies to use AI if it infringes on their privacy, even if those technologies improve their customer experiences. The survey involved more than 5,000 people in Australia, the UK, and the US.

In addition, nearly two-thirds (63%) of the survey's respondents said they're worried AI will make decisions about their lives without their knowledge.

Europe Setting the Bar

The EU has been leading the charge to meet consumer demand for digital privacy protections.

The EU's General Data Protection Regulation, or GDPR, went into effect in 2018 and vaulted digital privacy expectations to a higher level worldwide by ushering in new standards on a person's right to his or her own information, Ahmad said.

The EU's new privacy rules are taken seriously because of the potential fines for violations. Organisations can face fines up to the greater of €20 million ($22 million) or 4% of their annual global turnover if they are found to be out of compliance with the new privacy regulations. How data is stored, used, and protected is a focus of the GDPR, requiring companies to ensure their data collection and use policies and practices are in line with the privacy standard. Any business that uses personal data of persons in the EU to provide services, to sell goods, or to monitor their behaviour, even if those companies don't have an office in the EU, must comply with the rules. That requires tight control over how personal data is collected and processed.

Overseeing Data Privacy

Regardless of their size or scope, companies that are using these advanced technologies need to think through how their customer and client data is being protected and used, to ensure that people's privacy expectations aren't being violated unknowingly, Bernier said.

"While AI has been created post many privacy laws, the right to privacy and the way it has been defined and described and recognised by the courts does apply to AI," Bernier said.

Here are several ways to insert privacy concerns into management and corporate board discussion about AI:

Boards Can and Should Lead the Push for Privacy Protections
Boards can help hammer the point that any new technologies need to take security and privacy risks into account, Bernier said.

Board members don't have to understand the ins and outs of every new piece of technology, but they can make sure company management follows best practice in keeping consumers' data safe, she said. "They don't have to be the subject-matter expert, but they do need to know enough about the area to ask the right questions," she said.

Audit committee members can stress that they would like to see that routine checks of security protocols are being conducted and that plans are in place for any possible security breaches. (See the sidebar, "Tips for Board Members Dealing with AI," for more advice.)

Limit Yourself
Many companies will be better off collecting and storing fewer data points than stockpiling every bit of data available. That's because having large volumes of personal data, going back years, can lead to more problems if there's a security breach, Bernier said.

On top of that, having large data sets with hundreds of categories of information can make them unwieldy and make it hard to explain to consumers which variable led to a decision to decline them for a loan, turn them down for a job, or target a particular product towards them.

"You avoid excessive collection, and you get a logic model on a database that is manageable," Bernier said.

Companies should have routine schedules in place to examine what data they have on hand, with timetables to discard or thin out the information.

Think About Security from the Get-Go
Many companies, especially those just getting off the ground, focus on how to make their idea work and attract funding so they can get to the next point and scale up.

The thought of how to protect data and personalised information does not often become a primary concern in the first stages of a company's life, Ahmad said. Neglecting that puts businesses at a disadvantage.

By incorporating protections and best practice processes to routinely screen for issues, companies will be better off in the long term, he said. Corporate boards and top company officers should consider the data security piece whenever they are looking at AI for business solutions, Ahmad said.

"They really need to make sure that whatever solution or development is going to be used includes risk-based compliance," he said.

Inject Risk Awareness into Technology Discussions
Companies should be assessing their reputational, as well as monetary, risks of employing technologies that may create privacy concerns on an ongoing basis, said Atif Ansari, CPA (Canada), CGMA, the president and a founder of Canadian data analytics firm Piik Insights. That's not happening often enough, and those

in the C-suite need to ensure that any AI or other advanced technologies, like any other company networks, are protected from cyberattacks or breaches, Ansari said.

"It is incumbent on boards and executive management to consider the risks that a breach could pose," he said. "It should be on the radar."

That includes having discussions about what information is collected and shared, and how it is used to inform other business decisions.

His data analytics firm purposely doesn't use personally identifiable information in order to protect clients from inadvertent privacy disclosures, he said. Piik Insights works primarily with clients in the retail and restaurant sectors in North and South America.

For example, the company collects and uses only four digits of a customer's credit card to track their purchases and offer insight into consumer trends to its clients. "We can draw some analysis from this, but it doesn't personally identify any particular individuals," Ansari said.

Be Precise About Vendor Use of Data
Before signing up with any vendor, discuss how data provided by your company will be used and whether the vendor plans to use the data on other projects, Ahmad said.

Having those details spelled out in contracts, and making sure those contractual promises are kept, will go a long way towards making sure private information is protected.

"It's out of your control," Ahmad said about what happens once sensitive business information and customer data are handed off to a third party.

He also suggested considering worst-case scenarios and making sure those third parties have insurance policies that will cover the costs of any major cyberbreaches.

Make Sure the Analysis Does Not Go Too Far

Data thefts by cybercriminals aren't the only concern, Bernier said. Having the data go too far, and making conclusions that an individual is uncomfortable with, can raise other privacy concerns. She pointed to the now well-known case of the retailer Target, where the company's algorithm based on purchase history determined a teenager was pregnant before she was ready to talk to her family about it and sent coupons to her home.

Bernier recommended looking at what is done with customers' data and whether new privacy information is being created by linking up data and coming to a conclusion that could speak to a person's health, education, or other personal information.

Bringing Privacy to the Forefront

Corporate leaders are having more conversations about ways that advanced technologies and AI can affect individuals' privacy rights, and what to do about it, Ansari said.

"There's much more awareness, but there's still a lot more educating we need to do," he said.

Tips for Board Members Dealing with AI

- Encourage management to separate AI from analysis of other technology risks to break down the privatised information the technology creates and any risks that the data can be compromised.
- Make sure security protocols are followed by vendors long after contracts for services are signed. Encourage management to keep regular schedules to make sure technology partners are keeping their promises to protect personal information.
- Push management to comply with the most stringent set of privacy regulations, even if the company isn't currently in the EU or other markets with far-reaching requirements. That way, if the company does expand into those areas, it won't be an enormous burden to retrofit security protocols.

- Follow up with technology contractors to make sure security protocols are being followed. If an AI tool developed by a vendor is supposed to delete extraneous information, ask for verification that those deletions happen. The rule of thumb of privacy law expert Imran Ahmad, a Toronto-based lawyer with Blake, Cassels and Graydon who specialises in technology and cybersecurity issues, is to "trust but verify" that agreed-upon security practices are being followed.

Use Artificial Intelligence to Increase Food Production and Curb Poverty

Lisa Rabasca Roepe

Lisa Rabasca Roepe is a freelance journalist whose work has appeared in Fast Company, The Week, Quartz, and others.

The world population is expected to hit 9.6 billion by 2050, according to the United Nations, and experts warn that if scientists don't find more efficient ways to use and protect limited agricultural resources such as land, water, and energy, there could be a global food crisis. It's for this reason, eliminating world hunger is a top priority among the UN's Sustainable Development Goals, which the supranational organization hopes to attain by 2030.

Currently, one in nine people, or 795 million people, do not have enough food to lead a healthy, active life. For the third year in a row, world hunger has increased—in 2017, around 821 million people faced undernourishment from chronic food deprivation, the Food and Agriculture Organization of the United Nations reported.

As famine continues to increase to levels we haven't seen since a decade ago, it's becoming more critical to examine the way emerging technologies can curb this reoccurring problem. "We're in the midst of biggest [societal] shift since the Iron Age," said Elisabeth Mason, founding director at Stanford University's Poverty & Technology Lab. "How are we leveraging new technologies and skills to address these issues?"

Machine learning (ML) and artificial intelligence (AI) are among the new technologies leaders are relying on to help alleviate a global food crisis. Already, ML and AI technologies are predicting impoverished regions across the globe, using the data to find solutions to mitigate global hunger.

"How AI Can Help Fight Poverty," by Lisa Rabasca Roepe, Dell Inc., November 14, 2018. Reprinted by permission.

At Stanford, researchers are using these technologies to help humanitarian organizations measure the impact of their efforts, while Carnegie Mellon is tapping their potential to improve crops on a global scale. Here's a look at a few of these solutions in action.

An Abundant Food Source

Researchers at Carnegie Mellon are working with U.S. farmers to grow more high-value crops, such as grapes and apples, using machine learning, robots, and drones. The scientists plan to apply what they learn about fruit to breed staple crops. For example, "Any information we learn about growing high-value crops can be applied to growing sorghum," explained George A. Kantor, a senior systems scientist at Carnegie Mellon. Sorghum is common in the Sub-Saharan African and Asian regions and can withstand extreme heat and drought.

So how, exactly, does the technology work? The program uses a robot, sensors, and a high-quality camera to take photos of sorghum's grain head. On the back-end, AI technology looks at the photos and extracts information, such as the size of the grain head and the number and size of the seeds, then estimates the quality and ripeness of the crop.

The process allows crop breeders to compare over 1,000 varieties of sorghum and make better decisions about planting, cultivating, and harvesting. The ultimate goal, as the university's site states, is to help farmers develop "plants that produce more food on fewer acres with less water."

"We use robots and sensors and AI to improve the breeding process, and as a result the breeders end up with a new variety of sorghum that is higher yielding," Kantor explained. Eventually, instead of giving farmers in Africa and India advanced technology like robots and sensors, Carnegie Mellon researchers hope to hand these new seeds directly to local farmers to produce higher yielding crops. Rather than having to learn all new equipment, farmers would soon be able to plant seeds that just perform better.

Carnegie Mellon has also partnered with Clemson University to analyze plant growth. Plant breeders see potential with sorghum because, like corn, sorghum can be used as a source of grain for humans and livestock. However, the process to yield stronger sorghum is not instantaneous. "Plant breeding is slow and it takes multiple years to get a new product," Kantor said.

Predicting Areas of Poverty

At Stanford University, researchers from the Sustainability and Artificial Intelligence Lab are using machine learning and remote-sensing data to predict crop yields, specifically, as it relates to soybeans.

"If we have a model that works for U.S. soybeans, maybe we can train that model for areas with less data," said Marshall Burke, an assistant professor of earth system science at Stanford and a fellow at the Center on Food Security and the Environment. Researchers believe understanding crop yields will help farmers around the world make better planting decisions by increasing their ability to identify low-yield regions.

Stanford scientists are also working on locating areas of poverty, which can be difficult, as accurate and reliable data from impoverished regions is often scarce. To compensate for limited data, they are using machine learning technology to extract food scarcity information from high-resolution satellite imagery.

Researchers start out with data from household surveys that report agriculture and food insecurity on the ground, then use satellite imagery to model areas of poverty. When survey data isn't available, scientists can use the satellite imagery to predict areas of poverty, Burke explained. The ML algorithm can then use the imagery to determine whether areas have roads, farmland, and healthy vegetation.

Stanford is also working with a number of small NGOs and UN agencies, including the UN's World Food Programme, to determine if their relief efforts are having an impact. This research could help global organizations improve their humanitarian response to food shortages and distribute resources more effectively, Burke emphasized.

Disseminating Sustainable Solutions

At research institutions like Stanford and Carnegie Mellon, the focus isn't just trying to incubate solutions. According to Mason, it is also to build a network of practitioners across the United States, and to get technology companies to make investments in those solutions.

"AI and technology in general offer huge opportunities for us to universalize access to valuable information," she explained, "but also to target it in a way that can be more effective in opening up new opportunities."

AI Is Positive for Education Now and in the Future

William Goddard

William Goddard has a passion for technology and is the founder of IT Chronicles.

How do we use AI in education today?

In broad terms, AI or artificial intelligence is an attempt to create machines that can do things previously possible only through human perception, learning, or reasoning. While many researchers' ultimate goal is to develop AI capable of equaling and exceeding the full range of human cognition (a level known as artificial general intelligence, or AGI), we're probably still several decades away from such technology.

For the moment, artificial intelligence is narrow or weak, consisting of machines and systems with task-specific programming. Initially, so-called "expert systems" were developed by extensive programming rules into computers. More recently, however, AI has been advancing through the use of machine learning or ML—the observation and gathering of reams of data, identifying correlations that would not be immediately comprehensible to humans, and the use of those patterns to make decisions.

Artificial Intelligence in Education

This machine learning process has direct correlations to how humans learn. It should come as no surprise that these advances in technology are fueling the use of artificial intelligence in education.

With numerous stakeholders involved, there are plenty of avenues available for the use of artificial intelligence in education, and the market currently offers solutions for adults, children, tutors, and educational establishments. AI-based systems can analyze an

"AI in Education," by William Goddard, IT Chronicles, November 9, 2020. Reprinted by permission.

enormous amount of information, and the application of artificial intelligence in education covers a range that includes training, communications, administration, and resource management.

How Is Artificial Intelligence Used in Education?

For the most part, applications of artificial intelligence and machine learning in education take a virtual form, rather than being embodied like robots. There may be physical components involved, such as audio or visual sensors of the Internet of Things (IoT) that collect or observe environmental information. But artificial intelligence in education system applications generally manifests via digital software processing systems. This manifestation plays roles in education at various levels, including:

AI in Education: Developing Smart Content
AI systems can use the materials of a traditional syllabus to create customized textbooks for certain subjects. Such systems digitize this course material and create new learning interfaces to help students of all academic grades and ages.

Creating Personalized Learning Experiences
There's a quote by Albert Einstein that says that "Everybody is a genius, but if you judge a fish by its ability to climb a tree, it will live its whole life believing that it is stupid." No two students are the same, or learn in exactly the same way.

By providing customizable tutoring and studying support applications, AI can adapt educational frameworks to cater to individuals' needs according to their abilities, interests, and aptitude.

AI in Education: Expanding the Range of Education
In general, AI and digital technology are helping to eliminate boundaries and extend educational opportunities to learners throughout the world. Intelligent web search and recommendation engines can provide students with the information and resources they need to further their education. And platforms like the Massive Open Online Course or MOOC are making courses instantly

accessible to anyone with an internet connection. This is one of the significant benefits of artificial intelligence in education.

Facilitating Education Management and Administration

An education management information system or EMIS is an integrated group of information and documentation services to collect, store, process, analyze, and disseminate data for educational planning and management.

With the advancement of digital and machine learning technologies, these platforms are evolving into intelligent learning management systems (LMSs). From the masses of data collected through an EMIS, AI and machine learning algorithms can make data-driven decisions to improve school administration and education delivery.

AI in Education: Intelligent Tutoring and Learning

Intelligent tutoring system (ITS) and intelligent learning system (ILS) technologies provide AI-powered digital platforms that enable students to discover information for themselves. As well as acting as a medium for guided learning, they can also perform extensive diagnostics on student performance, maintaining a continuous model of their knowledge, skills, errors, and misconceptions—and providing recommendations or solutions to steer them along the path to progress.

10 Roles for Artificial Intelligence in Education

Artificial or machine intelligence can help students and teachers get more out of the educational experience by assuming a number of roles, including:

1. Automating Basic or Repetitive Activities: AI can currently automate grading for nearly all kinds of multiple-choice and fill-in-the-blank testing. As essay-grading software continues to evolve, this range of capabilities will expand.

2. Providing Personalized Learning Platforms: The growing numbers of adaptive learning programs, games, and software use AI to respond to the needs of individual students, laying greater emphasis on certain subjects, repeating things that students haven't mastered, and generally helping them to work at their own pace.
3. Identifying Gaps and Failings in the Curriculum: Systems like the electronic learning (eLearning) platform Coursera can give alerts if a large number of students submit the wrong answer to a homework assignment or have trouble with a particular topic.
4. Virtual Tutors: AI-powered tutoring systems are already helping students through basic mathematics, writing, and other subjects.
5. As a Feedback Loop for Students and Instructors: AI systems can monitor student progress and alert professors when there might be issues.
6. Providing New Ways to Interact with Information: As technologies evolve and integrate, students in the future may have increasingly immersive and diverse experiences doing research.
7. Creating a New Dynamic with Teachers: As AI becomes more integrated with the education system, teachers may supplement AI lessons, assisting students who are struggling, and providing human interaction and in-person experiences.
8. Reducing Academic and Social Pressures: With AI systems themselves often based on trial and error, they can facilitate this kind of learning in an environment where students don't feel pressured to compare themselves with their colleagues.
9. Providing New Avenues for Recruitment, Teaching, and Support: Smart data gathering powered by intelligent computer systems can provide information and

recommendations to enhance students' lives and the operations of educational institutions.
10. Changing the Dynamics of Learning: AI systems, software, and support, coupled with internet connectivity, enables students to learn from anywhere in the world at any time.

Artificial Intelligence in Higher Education: Current Uses and Future Applications

With high stakes riding on the outcomes of higher education—job prospects, funding for institutions, implications for the wider economy—there's a particular interest in acquiring the best tools, technologies, and talent to assure the success of establishments of higher learning. In this regard, artificial intelligence in higher education is contributing in a number of ways.

For recruiting and enrollment, some colleges and universities are using complex analytics systems based on machine learning that can calculate an individual's "demonstrated interest" in a particular course of study by tracking their interactions with institutional websites, social media posts, and email messages.

Enrollment analytics platforms can assist schools in determining which students they should reach out to, which aspects of campus life they should emphasize, and in the assessment of admissions applications.

Machine learning algorithms are also assisting higher education establishments in their marketing to prospective students, estimation of class sizes, the planning of the curriculum, and the allocation of resources.

For students, machine learning applications and platforms are helping to provide a number of support services. For example, some applications help students automatically schedule their course load, while others recommend courses, majors, and career paths. The tools can make recommendations based on how students with similar data profiles performed in the past.

AI software systems can use highly granular patterns of information and accurate assessments of student behavior. This analysis can help identify students who may be at risk—academically, socially, or financially. Institutions can use the information for determining which students merit financial aid or other interventions. However, these systems can raise concerns about individual privacy and autonomy.

From day to day, AI-based educational software can provide a personalized learning platform, with continuous and real-time assessments of student performance, and proactive recommendations of specific parts of a course for students to review, or additional resources for them to consult.

A striking instance of the use of artificial intelligence in science education comes from the annals of the Georgia Institute of Technology (Georgia Tech), where many of the students in a master's-level AI class were unaware that one of their teaching assistants (Jill Watson) wasn't human.

Jill was actually a digital entity, programmed to respond to a set of commonly repeated queries, and posted as one of the message board members. With a memory populated by tens of thousands of questions and answers from past semesters, Jill was one of the most effective teaching assistants the class had ever seen, answering questions with a success rate of 97 percent.

Looking ahead, higher institutions will face several challenges in their implementation and use of AI, as the technology evolves. Not least among these is the financial cost and the time that human workers will have to spend in training and curating data for advanced systems. Training will also be required for staff outside of IT teams who must be brought up to speed in the use of data and AI tools.

Other Artificial Intelligence Benefits in Education

Besides the advantages of using AI in education that we have already observed, there are other benefits, including:

- Continuous Access to Learning: With much of the AI technology-based online, education becomes an "anywhere, anytime" process, which may be undertaken more at the student's convenience.
- Access for Students with Special Needs: Innovative AI technologies are providing new ways of interacting for students with learning or physical disabilities and special needs.
- Increased Engagement: Personalized learning platforms with individualized course work, schedules, customized tasks, and interaction with digital technologies increase student engagement with the learning process, promoting retention and improved performance.
- Reduced Academic / Social Pressure: Material tailored to the needs of different learning groups enables students to proceed with their education without continually comparing themselves to others.

AI in Education: Some Potential Pitfalls

Of course, the use of AI in education isn't without complications and risks. Most notably:

- AI May Create Its Own Value System: By selecting the variables fed into admission, financial aid, or student information systems, AI tools are effectively creating rules about what matters in higher education. These criteria may not be the ideal ones.
- Institutions May Lose Sight or Control of Their Data: AI and ML systems rely on data management—and this is often contracted out to private companies that may be less directly accountable to the educational institution's stakeholders.
- Reliance on Data Risks Missing Out on Human Perception: Systems reliant on data and narrowly defined goals may miss the nuances and perceptions that would be seen by a human.

- There May Be Conflicts of Interest: Predictive analytics and early warning systems can promote student retention by drawing attention to those struggling. But they can also provide less benevolent institutions with ammunition for down-sizing their student populations.

Summary

This machine learning process has direct correlations to how humans learn. It should come as no surprise that these advances in technology are fueling the use of artificial intelligence in education. With numerous stakeholders involved, there are plenty of avenues available for the use of artificial intelligence in education, and the market currently offers solutions for adults, children, tutors, and educational establishments. AI-based systems can analyze an enormous amount of information, and the application of artificial intelligence in education covers a range that includes training, communications, administration, and resource management.

People Encounter Helpful AI Every Day

Sasha Reeves

Sasha Reeves is often a guest contributor on technology topics.

If you looked up the term "artificial intelligence" on Google and found your way to this article, you've used (and hopefully benefitted from) AI. If you've ever taken an Uber or had your phone auto-correct a misspelled word, you've used AI. Although it may not always be immediately obvious, artificial intelligence impacts nearly all aspects of our lives in a nearly uncountable number of ways. In this article, we'll take a look at eight examples of how artificial intelligence saves us time, money, and energy in our everyday life.

What Is Artificial Intelligence?

Before we can identify how artificial intelligence impacts our lives, it's helpful to know exactly what it is (and what it is not). The Oxford Dictionary defines artificial intelligence as:

> The theory and development of computer systems able to perform tasks that normally require human intelligence, such as visual perception, speech recognition, decision-making, and translation between languages.
>
> —The Oxford Dictionary of Phrase and Fable (2 ed.)

Essentially, artificial intelligence is the method by which a computer is able to act on data through statistical analysis, enabling it to understand, analyze, and learn from data through specifically designed algorithms. This is an automated process. Artificially intelligent machines can remember behavior patterns and adapt their responses to conform to those behaviors or encourage changes to them.

"8 Helpful Everyday Examples of Artificial Intelligence," by Sasha Reeves, IoT for All, August 10, 2020. Reprinted by permission.

The most important technologies that make up AI are machine learning (ML), deep learning, and natural language processing (NLP).

Machine learning is the process by which machines learn how better to respond based upon structured big data sets and ongoing feedback from humans and algorithms.

Deep learning is often thought to be a more advanced kind of ML because it learns through representation, but the data does not need to be structured.

Natural language processing (NLP) is a linguistic tool in computer science. It enables machines to read and interpret human language. NLP allows computers to translate human language into computer inputs.

8 Examples of Artificial Intelligence

Here is a list of eight examples of artificial intelligence that you're likely to come across on a daily basis.

1. Maps and Navigation

AI has drastically improved traveling. Instead of having to rely on printed maps or directions, you can now use Waze or Google, or Apple Maps on your phone and type in your destination.

So how does the application know where to go? And what's more, the optimal route, road barriers, and traffic congestions? Not too long ago, only satellite-based GPS was available, but now, artificial intelligence is being incorporated to give users a much more enhanced experience.

Using machine learning, the algorithms remember the edges of the buildings that it has learned, which allows for better visuals on the map, and recognition and understanding of house and building numbers. The application has also been taught to understand and identify changes in traffic flow so that it can recommend the route that avoids roadblocks and congestion.

2. Facial Detection and Recognition

Using virtual filters on our faces when taking pictures and using face ID for unlocking our phones are two examples of artificial intelligence that are now part of our daily lives. The former incorporates face detection, meaning any human face is identified. The latter uses face recognition through which a specific face is recognized. Facial recognition is also used for surveillance and security by government facilities and at airports.

3. Text Editors or Autocorrect

AI algorithms use machine learning, deep learning, and natural language processing to identify incorrect usage of language and suggest corrections in word processors, texting apps, and every other written medium, it seems. Linguists and computer scientists work together to teach machines grammar, just like you were taught at school. The algorithms are taught through high-quality language data so when you use a comma incorrectly, the editor will catch it.

4. Search and Recommendation Algorithms

When you want to watch a movie or shop online, have you noticed that the items suggested to you are often aligned with your interests or recent searches? These smart recommendation systems have learned your behavior and interests over time by following your online activity. The data is collected at the front end (from the user) and stored and analyzed through machine learning and deep learning. It is then able to predict your preferences, usually, and offer recommendations for things you might want to buy or listen to next.

5. Chatbots

As a customer, interacting with customer service can be time-consuming and stressful. For companies, it's an inefficient department that is typically expensive and hard to manage. One increasingly popular artificially intelligent solution to this is the use of AI chatbots. The programmed algorithms enable machines to answer frequently asked questions, take and track orders, and direct calls.

Chatbots are taught to impersonate the conversational styles of customer representatives through natural language processing (NLP). Advanced chatbots no longer require specific formats of inputs (e.g. yes/no questions). They can answer complex questions requiring detailed responses. In fact, if you give a bad rating for the response you get, the bot will identify the mistake it made and correct it for next time, ensuring maximum customer satisfaction.

6. Digital Assistants

When we have our hands full, we often resort to ordering digital assistants to perform tasks on our behalf. When you are driving, you might ask the assistant to call your mom (Don't text and drive, kids). A virtual assistant like Siri is an example of an AI that will access your contacts, identify the word "Mom," and call the number. These assistants use NLP, ML, statistical analysis, and algorithmic execution to decide what you are asking for and try to get it for you. Voice and image search work in much the same way.

7. Social Media

Social media applications are using the support of AI to monitor content, suggest connections, and serve advertisements to targeted users, among many other tasks to ensure that you stay invested and "plugged in."

AI algorithms can spot and swiftly take down problematic posts that violate terms and conditions through keyword identification and visual image recognition. The neural network architecture of deep learning is an important component of this process, but it doesn't stop there.

Social media companies know that their users are their product, so they use AI to connect those users to the advertisers and marketers that have identified their profiles as key targets. Social media AI also has the ability to understand the sort of content a user resonates with and suggests similar content to them.

8. E-Payments

Having to run to the bank for every transaction is an enormous waste of time and AI is playing a part in why you haven't been to a bank branch in 5 years. Banks are now leveraging artificial intelligence to facilitate customers by simplifying payment processes.

Intelligent algorithms have made it possible to make deposits, transfer money, and even open accounts from anywhere, leveraging AI for security, identity management, and privacy controls.

Even potential fraud can be detected by observing users' credit card spending patterns. This is also an example of artificial intelligence. The algorithms know what kind of products User X buys, when and from where they are typically bought, and in what price bracket they fall.

When there is an unusual activity that does not fit in with the user profile, the system can generate an alert or a prompt to verify transactions.

Final Takeaway

The examples of artificial intelligence that we have discussed not only serve as a source of entertainment but also provide many utilities that we use often. This technology is still developing and there are many more innovations yet to come.

CHAPTER 2

Will Artificial Intelligence Change What It Means to Be Human?

Overview: Advances in Technology Could Already Put Evolution into Hyperdrive

David Trippett

David Trippett is a senior lecturer at Cambridge University in the United Kingdom.

Biological evolution takes place over generations. But imagine if it could be expedited beyond the incremental change envisaged by Darwin to a matter of individual experience. Such things are dreamt of by so-called "transhumanists." Transhumanism has come to connote different things to different people, from a belief system to a cultural movement, a field of study to a technological fantasy. You can't get a degree in transhumanism, but you can subscribe to it, invest in it, research its actors, and act on its tenets.

So what is it? The term "transhumanism" gained widespread currency in 1990, following its formal inauguration by Max More, the CEO of Alcor Life Extension Foundation. It refers to an optimistic belief in the enhancement of the human condition through technology in all its forms. Its advocates believe in fundamentally enhancing the human condition through applied reason and a corporeal embrace of new technologies.

It is rooted in the belief that humans can and will be enhanced by the genetic engineering and information technology of today, as well as anticipated advances, such as bioengineering, artificial intelligence, and molecular nanotechnology. The result is an iteration of *Homo sapiens* enhanced or augmented, but still fundamentally human.

"Transhumanism: Advances in Technology Could Already Put Evolution into Hyperdrive—but Should They?" by David Trippett, The Conversation, March 28, 2018. https://theconversation.com/transhumanism-advances-in-technology-could-already-put-evolution-into-hyperdrive-but-should-they-92694. Licensed under CC BY-ND-4.0 International.

Evolution in Hyperdrive

The central premise of transhumanism, then, is that biological evolution will eventually be overtaken by advances in genetic, wearable and implantable technologies that artificially expedite the evolutionary process. This was the kernel of More's founding definition in 1990. Article two of the periodically updated, multi-authored "transhumanist declaration" continues to assert the point: "We favor morphological freedom—the right to modify and enhance one's body, cognition and emotions."

To date, areas to improve on include natural ageing (including, for die-hards, the cessation of "involuntary death") as well as physical, intellectual and psychological capacities. Some distinguished scientists, such as Hans Moravec and Raymond Kurzweil, even advocate a posthuman condition: the end of humanity's reliance on our congenital bodies by transforming "our frail version 1.0 human bodies into their far more durable and capable version 2.0 counterparts."

The push back against such unchecked optimism is emphatic. Some find the rhetoric distasteful in its assumptions about the desire for a prosthetic future.

And potential ethical problems, in particular, are raised. Tattoos, piercings and cosmetic surgery remain a matter of individual choice, and amputations a matter of medical necessity. But if augmented sensory capacity, for instance, were to become normative in a particular field, it might coerce others to make similar changes to their bodies in order to compete. As Isaiah Berlin once put it: "Freedom for the wolves has often meant death to the sheep."

Augmented Human Hearing

In order to really get to grips with the meaning of all this, though, an example is needed. Take the hypothetical augmentation of human hearing, something I am researching within a broader project on sound and materialism. Within discussions of transhumanism, ears are not typically among the sense organs figured for enhancement.

But human hearing is already being augmented. Algorithms for transposing auditory frequencies already exist (common to most speech processors in cochlear implants and hearing aids). Research into the regeneration of cilia hairs in the cochlear duct is also ongoing. Following this logic, augmenting unimpaired hearing need be no different, in principle, to correcting impaired hearing.

What next? Acoustic sound vibrations sit alongside the vast, inaudible electromagnetic spectrum, and various animals access different portions of this acoustic space, portions to which we—as humans—have no access. Could this change?

If it does, this may well alter the identity of sound itself. Speculations as to whether what is visible as light might under other circumstances be perceivable as sound have arisen at various points over the past two centuries. This raises heady questions about the very definition of sound. Must it be perceived by a human ear to constitute sound? By a sentient animal? Can a machine hear sufficiently to define sound beyond the human auditory range? What about aesthetics? Aesthetics itself—as the (human) study of the beautiful—may no longer even be applicable.

All Hypothetical?

The technologies for broaching such questions are arguably already at hand. Examples of auditory sense augmentation (broadly conceived) include Norbert Wiener's so-called "hearing glove," which stimulated the finger of a deaf person with electromagnetic vibrations; an implanted colour sensor that—for its colour-blind recipient, Neil Harbisson—converts the colour spectrum into sounds, including ultraviolet and infrared signals; and a cochlear implant that streams sounds wirelessly from Apple's mass market devices directly to the auditory nerve of its recipients.

The discussion is not entirely hypothetical, in other words. So what does all this mean?

There is a famous scene in the film *The Matrix* in which Morpheus asks Neo whether he wants to take the blue pill or the red pill. One returns him unawares to his life of total physical and mental enslavement within the simulation programme of the Matrix, the other gives him access to the real world with all its brutal challenges. But after experiencing this, he can never go back to life within the Matrix, and must survive outside it.

Advocates of transhumanism face a similar choice today. One option is to take advantage of the advances in nanotechnologies, genetic engineering and other medical sciences to enhance the biological and mental functioning of human beings (never to go back). The other is to legislate to prevent these artificial changes from becoming an entrenched part of humanity, with all the implied coercive bio-medicine that would entail for the species.

Of course, the reality of this debate is more complex. Holding our scepticism in abeyance, it still supersedes individual choice. Hence the question of agency remains: who should have the right to decide?

There Are Differences Between Human Intelligence and Artificial Intelligence

J. E. Korteling, G. C. van de Boer-Visschedijk, R. A. M. Blankendaal, R. C. Boonekamp, and A. R. Eikelboom

The authors are affiliated with Netherlands Organisation for Applied Scientific Research (TNO), Amsterdam, Netherlands.

[…]

Recent advances in information technology and in AI may allow for more coordination and integration between humans and technology. Therefore, quite some attention has been devoted to the development of human-aware AI, which aims at AI that adapts as a "team member" to the cognitive possibilities and limitations of the human team members. Also, metaphors like "mate," "partner," "alter ego," "Intelligent Collaborator," "buddy" and "mutual understanding" emphasize a high degree of collaboration, similarity, and equality in "hybrid teams." When human-aware AI partners operate like "human collaborators" they must be able to sense, understand, and react to a wide range of complex human behavioral qualities, like attention, motivation, emotion, creativity, planning, or argumentation (e.g. Krämer et al., 2012; van den Bosch and Bronkhorst, 2018; van den Bosch et al., 2019). Therefore these "AI partners" or "team mates" have to be endowed with human-like (or humanoid) cognitive abilities enabling mutual understanding and collaboration (i.e. "human awareness").

However, no matter how intelligent and autonomous AI agents become in certain respects, at least for the foreseeable future, they

J. E. Korteling, G. C. van de Boer-Visschedijk, R. A. M. Blankendaal, R. C. Boonekamp, and A. R. Eikelboom (2021), "Human Versus Artificial Intelligence," *Frontiers in Artifical Intellligence*, 4:622364. doi: 10.3389/frai.2021.622364. https://www.frontiersin.org/articles/10.3389/frai.2021.622364/full#h11. Licensed under CC BY-4.0 International.

probably will remain unconscious machines or special-purpose devices that support humans in specific, complex tasks. As digital machines they are equipped with a completely different operating system (digital vs biological) and with correspondingly different cognitive qualities and abilities than biological creatures, like humans and other animals (Moravec, 1988; Klein et al., 2004; Korteling et al., 2018a; Shneiderman, 2020a). In general, digital reasoning and problem-solving agents only compare very superficially to their biological counterparts (e.g. Boden, 2017; Shneiderman, 2020b). Keeping that in mind, it becomes more and more important that human professionals working with advanced AI systems (e.g. in military or policy making teams) develop a proper mental model about the different cognitive capacities of AI systems in relation to human cognition. This issue will become increasingly relevant when AI systems become more advanced and are deployed with higher degrees of autonomy. Therefore, the present paper tries to provide some more clarity and insight into the fundamental characteristics, differences and idiosyncrasies of human/biological and artificial/digital intelligences. In the final section, a global framework for constructing educational content on this "Intelligence Awareness" is introduced. This can be used for the development of education and training programs for humans who have to use or "collaborate with" advanced AI systems in the near and far future.

With the application of AI systems with increasing autonomy more and more researchers consider the necessity of vigorously addressing the real complex issues of "human-level intelligence" and more broadly artificial general intelligence, or AGI (e.g. Goertzel et al., 2014). Many different definitions of AGI have already been proposed (e.g. Russell and Norvig, 2014, for an overview). Many of them boil down to: technology containing or entailing (human-like) intelligence (e.g. Kurzweil, 1990). This is problematic. Most definitions use the term "intelligence" as an essential element of the definition itself, which makes the definition tautological. Second, the idea that AGI should be human-like seems unwarranted. At least in natural environments there are many other forms and manifestations of highly complex and intelligent behaviors that are very different from specific human cognitive abilities (see Grind, 1997, for an overview). Finally, like what is also frequently seen in the field of biology, these AGI definitions use human intelligence as a central basis or analogy for reasoning about the—less familiar—phenomenon of AGI (Coley and Tanner, 2012). Because of the many differences between the underlying substrate and architecture of biological and artificial intelligence this anthropocentric way of reasoning is probably unwarranted. For these reasons we propose a (non-anthropocentric) definition of "intelligence" as: "the capacity to realize complex goals" (Tegmark, 2017). These goals may pertain to narrow, restricted tasks (narrow AI) or to broad task domains (AGI). Building on this definition, and on a definition of AGI proposed by Bieger et al. (2014) and one of Grind (1997), we define AGI here as: "Non-biological capacities to autonomously and efficiently achieve complex goals in a wide range of environments." AGI systems should be able to identify and extract the most important features for their operation and learning process automatically and efficiently over a broad range of tasks and contexts. Relevant AGI research differs from the ordinary AI research by addressing the versatility and wholeness of intelligence, and by carrying out the engineering practice according to a system comparable to the human mind in a certain sense (Bieger et al., 2014).

It will be fascinating to create copies of ourselves that can learn iteratively by interaction with partners and thus become able to collaborate on the basis of common goals and mutual understanding and adaptation (e.g. Bradshaw et al., 2012; Johnson et al., 2014). This would be very useful, for example when a high degree of social intelligence of AI will contribute to more adequate interactions with humans, for example in health care or for entertainment purposes (Wyrobek et al., 2008). True collaboration on the basis of common goals and mutual understanding necessarily implies some form of humanoid general intelligence. For the time being, this remains a goal on a far-off horizon. In the present paper we argue why for most applications it also may not be very practical or necessary (and probably a bit misleading) to vigorously aim or to anticipate on systems possessing "human-like" AGI or "human-like" abilities or qualities. The fact that humans possess general intelligence does not imply that new inorganic forms of general intelligence should comply to the criteria of human intelligence. In this connection, the present paper addresses the way we think about (natural and artificial) intelligence in relation to the most probable potentials (and real upcoming issues) of AI in the short- and mid-term future. This will provide food for thought in anticipation of a future that is difficult to predict for a field as dynamic as AI.

Implicit in our aspiration of constructing AGI systems possessing humanoid intelligence is the premise that human (general) intelligence is the "real" form of intelligence. This is even already implicitly articulated in the term "Artificial Intelligence," as if it were not entirely real, i.e., real like non-artificial (biological) intelligence. Indeed, as humans we know ourselves as the entities with the highest intelligence ever observed in the Universe. And as an extension of this, we like to see ourselves as rational beings who are able to solve a wide range of complex problems under all kinds of circumstances using our experience and intuition, supplemented by the rules of logic, decision analysis and statistics. It is therefore not surprising that we have some difficulty to accept the idea that we might be a bit less smart than we keep on telling ourselves, i.e., "the next insult for humanity" (van Belkom, 2019). This goes as far that the rapid progress in the field of artificial intelligence is accompanied by a recurring redefinition of what should be considered "real (general) intelligence." The conceptualization of intelligence, that is, the ability to autonomously and efficiently achieve complex goals, is then continuously adjusted and further restricted to: "those things that only humans can do." In line with this, AI is then defined as "the study of how to make computers do things at which, at the moment, people are better" (Rich and Knight, 1991; Rich et al., 2009). This includes thinking of creative solutions, flexibly using contextual and background information, the use of intuition and feeling, the ability to really "think and understand," or the inclusion of emotion in an (ethical) consideration. These are then cited as the specific elements of real intelligence (e.g. Bergstein, 2017). For instance, Facebook's director of AI and a spokesman in the field, Yann LeCun, mentioned at a conference at MIT on the Future of Work that machines are still far from having "the essence of intelligence." That includes the ability to understand the physical world well enough to make predictions about basic aspects of it—to observe one thing and then use background knowledge to figure out what other things must also be true. Another way of saying this is that machines don't have common sense (Bergstein, 2017), like submarines that cannot swim (van Belkom, 2019). When exclusive human capacities become our pivotal navigation points on the horizon we may miss some significant problems that may need our attention first.

[...]

We Are Probably Not so Smart as We Think

How intelligent are we actually? The answer to that question is determined to a large extent by the perspective from which this issue is viewed, and thus by the measures and criteria for intelligence that is chosen. For example, we could compare the nature and capacities of human intelligence with other animal species. In that case we appear highly intelligent. Thanks to our enormous learning capacity, we have by far the most extensive arsenal of cognitive abilities to autonomously solve complex problems and achieve complex objectives. This way we can solve a huge variety of arithmetic, conceptual, spatial, economic, socio-organizational, political, etc. problems. The primates—which differ only slightly from us in genetic terms—are far behind us in that respect. We can therefore legitimately qualify humans, as compared to other animal species that we know, as highly intelligent.

Limited Cognitive Capacity

However, we can also look beyond this "relative interspecies perspective" and try to qualify our intelligence in more absolute terms, i.e., using a scale ranging from zero to what is physically possible. For example, we could view the computational capacity of a human brain as a physical system (Bostrom, 2014; Tegmark, 2017). The prevailing notion in this respect among AI scientists is that intelligence is ultimately a matter of information and computation, and (thus) not of flesh and blood and carbon atoms. In principle, there is no physical law preventing that physical systems (consisting of quarks and atoms, like our brain) can be built with a much greater computing power and intelligence than the human brain. This would imply that there is no insurmountable physical reason why machines one day cannot become much more intelligent than ourselves in all possible respects (Tegmark, 2017). Our intelligence is therefore relatively high compared to other animals, but in absolute terms it may be very limited in its physical computing capacity, albeit only by the limited size of our brain and its maximal possible number of neurons and glia cells (e.g. Kahle, 1979).

To further define and assess our own (biological) intelligence, we can also discuss the evolution and nature of our biological thinking abilities. As a biological neural network of flesh and blood, necessary for survival, our brain has undergone an evolutionary optimization process of more than a billion years. In this extended period, it developed into a highly effective and efficient system for regulating essential biological functions and performing perceptive-motor and pattern-recognition tasks, such as gathering food, fighting and flighting, and mating. Almost during our entire evolution, the neural networks of our brain have been further optimized for these basic biological and perceptual motor processes that also lie at the basis of our daily practical skills, like cooking, gardening, or household jobs. Possibly because of the resulting proficiency for these kinds of tasks we may forget that these processes are characterized by extremely high computational complexity (e.g. Moravec, 1988). For example, when we tie our shoelaces, many millions of signals flow in and out through a large number of different sensor systems, from tendon bodies and muscle spindles in our extremities to our retina, otolithic organs and semi-circular channels in the head (e.g. Brodal, 1981). This enormous amount of information from many different perceptual-motor systems is continuously, parallel, effortless and even without conscious attention, processed in the neural networks of our brain (Minsky, 1986; Moravec, 1988; Grind, 1997). In order to achieve this, the brain has a number of universal (inherent) working mechanisms, such as association and associative learning (Shatz, 1992; Bar, 2007), potentiation and facilitation (Katz and Miledi, 1968; Bao et al., 1997), saturation and lateral inhibition (Isaacson and Scanziani, 2011; Korteling et al., 2018a).

These kinds of basic biological and perceptual-motor capacities have been developed and set down over many millions of years. Much later in our evolution—actually only very recently—our cognitive abilities and rational functions have started to develop. These cognitive abilities, or capacities, are probably less than 100 thousand years old, which may be qualified as "embryonal" on the time scale of evolution (e.g. Petraglia and Korisettar, 1998; McBrearty and Brooks, 2000; Henshilwood and Marean, 2003). In addition, this very thin layer of human achievement has necessarily been built on these "ancient" neural intelligence for essential survival functions. So, our "higher" cognitive capacities are developed from and with these (neuro) biological regulation mechanisms (Damasio, 1994; Korteling and Toet, 2020). As a result, it should not be a surprise that the capacities of our brain for performing these recent cognitive functions are still rather limited.

[…]

Ingrained Cognitive Biases

Our limited processing capacity for cognitive tasks is not the only factor determining our cognitive intelligence. Except for an overall limited processing capacity, human cognitive information processing shows systematic distortions. These are manifested in many cognitive biases (Tversky and Kahneman, 1973; Tversky and Kahneman, 1974). Cognitive biases are systematic, universally occurring tendencies, inclinations, or dispositions that skew or distort information processes in ways that make their outcome inaccurate, suboptimal, or simply wrong (e.g. Lichtenstein and Slovic, 1971; Tversky and Kahneman, 1981). Many biases occur in virtually the same way in many different decision situations (Shafir and LeBoeuf, 2002; Kahneman, 2011; Toet et al., 2016). The literature provides descriptions and demonstrations of over 200 biases. These tendencies are largely implicit and unconscious and feel quite naturally and self/evident when we are aware of these cognitive inclinations (Pronin et al., 2002; Risen, 2015; Korteling et al., 2018b). That is why they are often termed "intuitive" (Kahneman and Klein, 2009) or "irrational" (Shafir and LeBoeuf, 2002). Biased reasoning can result in quite acceptable outcomes in natural or everyday situations, especially when the time cost of reasoning is taken into account (Simon, 1955; Gigerenzer and Gaissmaier, 2011). However, people often deviate from rationality and/or the tenets of logic, calculation, and probability in inadvisable ways (Tversky and Kahneman, 1974; Shafir and LeBoeuf, 2002) leading to suboptimal decisions in terms of invested time and effort (costs) given the available information and expected benefits.

Biases are largely caused by inherent (or structural) characteristics and mechanisms of the brain as a neural network (Korteling et al., 2018a; Korteling and Toet, 2020). Basically, these mechanisms—such as association, facilitation, adaptation, or lateral inhibition—result in a modification of the original or available data and its processing (e.g. weighting its importance). For instance, lateral inhibition is a universal neural process resulting in the magnification of differences in neural activity (contrast enhancement), which is very useful for perceptual-motor functions, maintaining physical integrity and allostasis (i.e. biological survival functions). For these functions our nervous system has been optimized for millions of years. However, "higher" cognitive functions, like conceptual thinking, probability reasoning or calculation, have been developed only very recently in evolution. These functions are probably less than 100 thousand years old, and may, therefore, be qualified as "embryonal" on the time scale of evolution (e.g. McBrearty and Brooks, 2000; Henshilwood and Marean, 2003; Petraglia and Korisettar, 2003). In addition, evolution could not develop these new cognitive functions from scratch, but instead had to build this embryonal, and thin layer of human achievement from its "ancient" neural heritage for the essential biological survival functions (Moravec, 1988). Since cognitive functions typically require exact calculation and proper weighting of data, data transformations—like lateral inhibition—may easily lead to systematic distortions (i.e. biases) in cognitive information processing. Examples of the large number of biases caused by the inherent properties of biological neural networks are: Anchoring bias (biasing decisions toward previously acquired information, Furnham and Boo, 2011; Tversky and Kahneman, 1973; Tversky and Kahneman, 1974), the Hindsight bias (the tendency to erroneously perceive events as inevitable or more likely once they have occurred, Hoffrage et al., 2000; Roese and Vohs, 2012), the Availability bias (judging the frequency, importance, or likelihood of an event by the ease with which relevant instances come to mind, Tversky and Kahnemann, 1973; Tversky and Kahneman, 1974), and the Confirmation bias (the tendency to select, interpret, and remember information in a way that confirms one's preconceptions, views, and expectations, Nickerson, 1998). In addition to these inherent (structural) limitations of (biological) neural networks, biases may also originate from functional evolutionary principles promoting the survival of our ancestors who, as hunter-gatherers, lived in small, close-knit groups (Haselton et al., 2005; Tooby and Cosmides, 2005). Cognitive biases can be caused by a mismatch between evolutionarily rationalized "heuristics" ("evolutionary rationality": Haselton et al., 2009) and the current context or environment (Tooby and Cosmides, 2005). In this view, the same heuristics that optimized the chances of survival of our ancestors in their (natural) environment can lead to maladaptive (biased) behavior when they are used in our current (artificial) settings. Biases that have been considered as examples of this kind of mismatch are the Action bias (preferring action even when there is no rational justification to do this, Baron and Ritov, 2004; Patt and Zeckhauser, 2000), Social proof (the tendency to mirror or copy the actions and opinions of others, Cialdini, 1984), the Tragedy of the commons (prioritizing personal interests over the common good of the community, Hardin, 1968), and the Ingroup bias (favoring one's own group above that of others, Taylor and Doria, 1981).

This hard-wired (neurally inherent and/or evolutionary ingrained) character of biased thinking makes it unlikely that simple and straightforward methods like training interventions or awareness courses will be very effective to ameliorate biases. This difficulty of bias mitigation seems indeed supported by the literature (Korteling et al., 2021).

General Intelligence Is Not the Same as Human-like Intelligence

We often think and deliberate about intelligence with an anthropocentric conception of our own intelligence in mind as an obvious and unambiguous reference. We tend to use this conception as a basis for reasoning about other, less familiar phenomena of intelligence, such as other forms of biological and artificial intelligence (Coley and Tanner, 2012). This may lead to fascinating questions and ideas. An example is the discussion about how and when the point of "intelligence at human level" will be achieved. For instance, Ackermann (2018) writes: "Before reaching superintelligence, general AI means that a machine will have the same cognitive capabilities as a human being." So, researchers deliberate extensively about the point in time when we will reach general AI (e.g., Goertzel, 2007; Müller and Bostrom, 2016). We suppose that these kinds of questions are not quite on target. There are (in principle) many different possible types of (general) intelligence conceivable of which human-like intelligence is just one of those. This means, for example, that the development of AI is determined by the constraint of physics and technology, and not by those of biological evolution. So, just as the intelligence of a hypothetical extraterrestrial visitor of our planet earth is likely to have a different (in-)organic structure with different characteristics, strengths, and weaknesses, than the human residents this will also apply to artificial forms of (general) intelligence.

[…]

These kinds of differences in basic structure, speed, connectivity, updatability, scalability, and energy consumption will necessarily also lead to different qualities and limitations between human and artificial intelligence. Our response speed to simple stimuli is, for example, many thousands of times slower than that of artificial systems. Computer systems can very easily be connected directly to each other and as such can be part of one integrated system. This means that AI systems do not have to be seen as individual entities that can easily work alongside each other or have mutual misunderstandings. And if two AI systems are engaged in a task then they run a minimal risk to make a mistake because of miscommunications (think of autonomous vehicles approaching a crossroad). After all, they are intrinsically connected parts of the same system and the same algorithm (Gerla et al., 2014).

[…]

The Impact of Multiple Narrow AI Technology
AGI as the Holy Grail

AGI, like human general intelligence, would have many obvious advantages, compared to narrow (limited, weak, specialized) AI. An AGI system would be much more flexible and adaptive. On the basis of generic training and reasoning processes it would understand autonomously how multiple problems in all kinds of different domains can be solved in relation to their context (e.g. Kurzweil, 2005). AGI systems also require far fewer human interventions to accommodate the various loose ends among partial elements, facets, and perspectives in complex situations.

AGI would really understand problems and is capable to view them from different perspectives (as people—ideally—also can do). A characteristic of the current (narrow) AI tools is that they are skilled in a very specific task, where they can often perform at superhuman levels (e.g. Goertzel, 2007; Silver et al., 2017). These specific tasks have been well-defined and structured. Narrow AI systems are less suitable, or totally unsuitable, for tasks or task environments that offer little structure, consistency, rules or guidance, in which all sorts of unexpected, rare or uncommon events (e.g. emergencies) may occur. Knowing and following fixed procedures usually does not lead to proper solutions in these varying circumstances. In the context of (unforeseen) changes in goals or circumstances, the adequacy of current AI is considerably reduced because it cannot reason from a general perspective and adapt accordingly (Lake et al., 2017; Horowitz, 2018). As with narrow AI systems, people are then needed to supervise on these deviations in order to enable flexible and adaptive system performance. Therefore the quest of AGI may be considered as looking for a kind of holy grail.

Multiple Narrow AI Is Most Relevant Now!
The potential high prospects of AGI, however, do not imply that AGI will be the most crucial factor in future AI R&D, at least for the short- and mid-term. When reflecting on the great potential benefits of general intelligence, we tend to consider narrow AI applications as separate entities that can very well be outperformed by a broader AGI that presumably can deal with everything. But just as our modern world has evolved rapidly through a diversity of specific (limited) technological innovations, at the system level the total and wide range of emerging AI applications will also have a groundbreaking technological and societal impact (Peeters et al., 2020). This will be all the more relevant for the future world of big data, in which everything is connected to everything through the Internet of Things. So, it will be much more profitable and beneficial to develop and build (non-human-like) AI variants that will excel in areas where people are inherently limited. It seems not

too far-fetched to suppose that the multiple variants of narrow AI applications also gradually get more broadly interconnected. In this way, a development toward an ever broader realm of integrated AI applications may be expected. In addition, it is already possible to train a language model AI (Generative Pre-trained Transformer 3, GPT-3) with a gigantic dataset and then have it learn various tasks based on a handful of examples—one or few-shot learning. GPT-3 (developed by OpenAI) can do this with language-related tasks, but there is no reason why this should not be possible with image and sound, or with combinations of these three (Brown, 2020).

Besides, the moravec paradox implies that the development of AI "partners" with many kinds of human (-level) qualities will be very difficult to obtain, whereas their added value, (i.e. beyond the boundaries of human capabilities) will be relatively low. The most fruitful AI applications will mainly involve supplementing human constraints and limitations. Given the present incentives for competitive technological progress, multiple forms of (connected) narrow AI systems will be the major driver of AI impact on our society for short- and mid-term. For the near future, this may imply that AI applications will remain very different from, and in many aspects almost incomparable with, human agents. This is likely to be true even if the hypothetical match of artificial general intelligence (AGI) with human cognition were to be achieved in the future in the longer term. Intelligence is a multi-dimensional (quantitative, qualitative) concept. All dimensions of AI unfold and grow along their own different path with their own dynamics. Therefore, over time an increasing number of specific (narrow) AI capacities may gradually match, overtake and transcend human cognitive capacities. Given the enormous advantages of AI, for example in the field of data availability and data processing capacities, the realization of AGI probably would at the same time outclass human intelligence in many ways. Which implies that the hypothetical point of time of matching human- and artificial cognitive capacities, i.e. human-level AGI, will probably be hard to define in a meaningful way (Goertzel, 2007).

So when AI will truly understand us as a "friend," "partner," "alter ego" or "buddy," as we do when we collaborate with other humans as humans, it will surpass us in many areas at the same Moravec (1998) time. It will have a completely different profile of capacities and abilities and thus it will not be easy to really understand the way it "thinks" and comes to its decisions. In the meantime, however, as the capacities of robots expand and move from simple tools to more integrated systems, it is important to calibrate our expectations and perceptions toward robots appropriately. So, we will have to enhance our awareness and insight concerning the continuous development and progression of multiple forms of (integrated) AI systems. This concerns for example the multi-facetted nature of intelligence. Different kind of agents may have different combinations of intelligences of very different levels. An agent with general intelligence may for example be endowed with excellent abilities on the area of image recognition and navigation, calculation, and logical reasoning while at the same time being dull on the area of social interaction and goal-oriented problem solving. This awareness of the multi-dimensional nature of intelligence also concerns the way we have to deal with (and capitalize on) anthropomorphism. That is the human tendency in human-robot interaction to characterize non-human artifacts that superficially look similar to us as possessing human-like traits, emotions, and intentions (e.g., Kiesler and Hinds, 2004; Fink, 2012; Haring et al., 2018). Insight into these human factors issues is crucial to optimize the utility, performance and safety of human-AI systems (Peeters et al., 2020).

From this perspective, the question whether or not "AGI at the human level" will be realized is not the most relevant question for the time being. According to most AI scientists, this will certainly happen, and the key question is not IF this will happen, but WHEN (e.g., Müller and Bostrom, 2016). At a system level, however, multiple narrow AI applications are likely to overtake human intelligence in an increasingly wide range of areas.

[…]

Robots Will Be Human-Like

Progressive Automations Inc.

Progressive Automations Inc. is a company that specializes in automation.

Modern robotic technology is an extremely dynamic branch of science that develops rapidly and brings interesting breakthrough innovations almost every year. Robots are commonly used in our day to day life to do all sorts of things. Human-like machines are implemented everywhere from medicine to householding and manufacturing. These robots can vary in their characteristics, functions, and outlook; some of them range from rather primal and simple in construction to very complicated and almost indistinct from humans. As of now, creating robots that look like humans is the main goal of robotics all around the world.

Robots in Real Life

Robots have quite a wide scope of application. They can be useful in almost every possible area of life.

Robots in Medicine

Robots have been used for treating disabled people for a long time now. In particular, the process of production of up-to-date prostheses and artificial limbs is highly dependent on robotics. The prostheses that meet today's high standards of medicine are movable, mobile, flexible, and serve as qualitative replacements for lost limbs. Many of these artificial body parts are manipulated directly by the electronic impulses of the patient's body. But this isn't all that robotics can offer—robots can be found in surgical wards, as well as many educational institutions for doctors, and the range of their possible applications only grows.

"Most Human-Like Robots," Progressive Automations Inc., December 23, 2020. Reprinted by permission.

Robots in the Aerospace Industry

No one will argue that robots should find their place in future space exploration missions as an indispensable part of outer space-traveling crews. And even if we resort to history, we can already see that most of the space explorations were actually conducted with the help of robots, so this idea is not something new, either. We have numerous Moon Rovers, Mars Rovers, and avatar-robots working in outer space for us. But the diversity of these machines doesn't end on that, there are actually a lot of different types of robots designed specifically for the extreme conditions of space. They can perform tasks inaccessible to humans or those that are just way too dangerous.

Independent Robotic Security Systems

The implementation of modern robots in the area of life concerning the security of humans has proved to be highly valuable. For example, there are robots that can locate and identify potential fire dangers and eliminate them, preventing catastrophes before they even happen.

And we shouldn't dismiss the robots engaged in military services. They are used to conduct soldier training—despite not being designed reminiscent of human beings like other high-end robots, they are perfect at reproducing the behavior and reactions of a real person. Using these machines, military leadership creates situations and conditions that are extremely similar to real hostilities.

Robots can perform spying operations on different objects, not raising any attention to themselves. This way, they take some weight off the shoulders of law enforcement officers, allowing them to engage in other important tasks.

Robot Appliances in Households and Manufacturing

It is hard to imagine modern production facilities and manufacturing plants without all the robots that they are using. Usually, robots are responsible for the majority of work done on those plants. We talk about a massive number of operations requiring utmost accuracy of movement and numerous reiterations boiling down to a single continuous high-speed process—an impossible scenario for a human. Implementing robots can increase the production rates of entire manufacturing sectors, freeing up human employees and allowing them to take over the other important objectives of the enterprise.

Robots are also often used in households to help the owners with their cooking, cleaning, and other routine work around the house. The most frequent are the robot-vacuum cleaners and robot-lawnmowers but there are also other machines that perform much more complicated tasks.

Robots in the Entertainment Industry

There are also robots engineered for the sole purpose of entertaining humans. Those are mostly children's toys that can sing a song or make a few dance movements, radio remote control car toys, and some interactive games. The market even has the so-called real robots that can communicate using scripted speech patterns, so it seems that not all of them are made for children after all. Though, the only difference between the models for adult users and the ones for kids is the design and size of a toy.

Famous Robots

Hundreds of companies all across the world produce massive numbers of different kinds of robotic mechanisms every day. To make your robot stand out among the others, it has to be a real gem in the robotics realm. Here are some examples of such works, famous for their unique functions.

Asimo

This Japanese robot was developed by the Honda company. Honda has actually been quite an active player in the robotics realm since the 1980s. Their specialists were able to construct a robot that successfully occupied all the attention of the entire robotics sector of science. With the weight of 50 kilos and reaching 1.5 meters in height, this fella could move on its own, avoiding different obstacles and executing additional scripted programs of actions such as serving a cup of coffee to a human.

VGo

This robot is controlled over a Wi-Fi network and is a so-called telepresence device. It can communicate via a microphone, as well as hear and see its surroundings, which allows it to move around freely. The mechanism was developed specifically to serve disabled people who are unable to do everyday duties on their own. If you were to connect this robot to certain devices, it could be used as an independent portable camera. With the help of this technology, children who are currently homeschooling can listen to the school lessons from their homes.

Roboy

This robot was developed by specialists from Zurich. It has flexible tendons, which creates a more anthropomorphic look. The designers tried to put together a face that would express emotions, along with soft material used as a skin, and it looks decent. This realistic robot was initially elaborated as a companion for lonely people who require some attention from outside.

Kuratas

This giant robot weighs about 4.5 tons and reaches a fascinating 4 meters in height. Its movement is manipulated by a controller who is placed in a special cabin—but it can also be controlled remotely. The design of the robot immediately reminds one of an anime character, a rather expected solution coming from the Japanese engineers.

iCub

One of the most advanced robots was built in Italy. From the outside, it is very reminiscent of a real person already. But if you take a deeper look at its functionality you will find that this robot is capable of recalling its own name and names of those who it speaks to, as well as the names of different objects. Moreover, it can adapt to its surroundings and move freely by itself.

Rollin' Justin
Rollin' Justin was developed by the German Aerospace Center. It is controlled via telepresence technology, with the program being remotely directed by an operator through the monitor screen. The robot comes in different model modifications, including the straight standing version.

What Are Robots Made Of?
There are several different materials used to build a robot. The most frequent options include:

- Aluminum
- Steel
- Plexiglass
- Plastic
- Carbon fiber
- Synthetic rubber (used to represent artificial skin and muscle tissue in realistic robots)

The choice of materials depends directly on the specific functions of a robot and its working conditions. The worse the environment that a robot is going to be working in—the stronger and tougher the materials should be to resist wearing down.

Synthetic Robots
Construction of synthetic or "soft" robots is one of the high-priority directions of the modern sector of robotics working on realistic robots. These synthetic humanoids are generally defined by their smooth surface intended to resemble human skin. They are built on the bionic principles and designed specifically for use on working places in tight contact with humans due to much lesser possibilities of trauma as opposed to working close to an analogous mechanism of metal and hard plastic. Thanks to the artificial skin and muscle

tissue used in the construction of the robots, the final result is highly anthropomorphic!

Robots vs Androids

Android robots are humanoid-robots that look exactly like real humans or at least designed to resemble them in their outlook and actions. Such models are currently in active development by many famous leading robot manufacturers all around the world. A normal robot, on the other hand, usually looks less like a human and has much fewer functions.

Robots have proved to be useful in situations when you need to get some uncomplicated repetitive work done fast or when you need to perform high accuracy operations or lift heavy objects. An android robot is a type of robot you can assign operations requiring intellect to. It can deal with tasks such as communicating or providing help directly to a person in different areas of life—its capacities are much more diverse as compared to ordinary robots designed to do strictly scripted work.

Companion Robot for Personal Interaction

Robot-companions are designed to provide useful socially appropriate interactions in the real-world environment. The first and most important task of such a robot is to serve humans and help them get their routine done. Most and foremost, robot-companions, as follows from their name, were designed for people who require someone to be there for them, people with specific needs: elderly, disabled or autistic people. These robots can be of great help to those in need.

In recent years, human-looking robots are becoming more and more popular among enthusiasts and those curious enough to pay attention to the funny little robots sharing some human features and naturally arousing sympathy. The main function of these robots is to keep up conversations with anybody talking to them. They are a great means of communication for rich people who are, nonetheless, lonely and in acute need to share their thoughts.

There are also robot-companions that look like home pets available at the robotic market. Though, more often than not, they serve only as morale boosters having little practical application.

What Is a Humanoid?

Robot-humanoids are devices that at least partly resemble humans on the outside, even if they don't have any special functions to back it up. Most of them are equipped with hands, legs, head, and even a face on it. Despite the fact that the concept of these robots comes from way back in the day, the most prominent achievements in this sphere were made in the last decade.

A real humanoid should not only resemble a person from outside but also has to have a range of autonomous abilities such as moving freely, listening, and speaking on its own. And perhaps the most important feature of a modern humanoid is the ability to refine its skills during the process of performing certain actions, just like a real person would do.

How to Make a Humanoid Robot

Versatile human-like robots are extremely complicated machines that require quite a lot of knowledge and experience to put together a decent model. The construction of such a robot must include the musculoskeletal system, mechanical limbs, systems of sound and environment recognition, as well as a neural-network capable of gathering and processing information from outside. The elements of appearance such as its skin and hair, are not necessary and serve more as an aesthetic part of the whole mechanism but can be dismissed.

Build a Humanoid Robot

Most of the robots share some similarities. These include, first of all, agile bodies allowing them to move around freely. Some models have more movable elements in their bodies, some can move only the main parts, but it has to be enough to avoid a robot from tripping over obstacles. As a rule, the body parts are made of metal and plastic interconnected with a system of joints just like human bones are.

Robots move around by activating the right drivers at the appropriate time harmoniously. Keeping the robots from falling nose first on the ground is harder than it may seem, provided that it has to remain on its feet the whole time. Some use electric motors, linear actuators, and solenoids, others use hydraulic pneumatic systems.

A robot has to have a charging source to stay active. Most robots have integral batteries that can be charged from a power grid. Hydraulic models have pumps for creating tension, while pneumatic ones use tanks with compressed air or special compressors to do the job.

All the robot's motors connect to power grids that charge electronic engines and solenoids directly. Everything connected to the grid is controlled by a computer that activates the necessary details and valves. The behavior of a robot can be altered easily just by changing the program it has installed in its system.

Although some models have the ability to hear sounds, see images, smell objects, and feel touching on their sensors, not every robot has such systems. But the absolute must for a modern robot is the ability to move—otherwise, it can not be considered a robot.

Future Humanoid Robots

At the moment, robotics has several different routes of development it can follow that, according to the forecasts of the specialists in this realm, can become the main trends in the future of robotics as a whole. These include:

- Using new materials, namely galliot nitrate (in the production of transistors) and graphite (in the production of drive units);
- Using new sources of energy as well as the technology of its gathering and storing;
- Robots interacting with humans, for example, to manage unmanned aircraft traffic;
- Helping robots with navigation in extreme conditions;

- Increasing neural-network efficiency by boosting up the complexity of its structure and architecture;
- Creating possibilities to study algorithmic procedures and usage of cloud services for mass education to robots;
- Using artificial intelligence for perfecting movement patterns;
- Implementing robots in different areas of human life to increase their financial profitability;
- Concentrating on developing the abilities and quality of sensors.

Bearing in mind that the developments in all of the above-listed fields are currently being made, it is a fact that in the nearest future the world of robotics will surprise us with some high-class robots, unlike everything we have seen so far.

At the moment, human-like robots already do a lot of work in numerous areas of life. And in the future, those numbers only seem to grow. The specter of tasks robots can be used for will only increase along with their abilities. They will be used to do jobs dangerous for humans, for jobs that require scrupulously reiterating tasks, catering for elderly and children, educating children and adults, and many other things that can be completed by robots. And while doing all these jobs for humans, robots will be developing even further, which opens up more and more prospects for the future of robotics.

Human Intelligence Is Superior to AI in Specific Ways

Ben Dickson

Ben Dickson is a software engineer who writes about technology, politics, and business.

These days, it's easy to believe arguments that artificial intelligence has become as smart as the human mind—if not smarter. Google released a speaking AI that dupes its conversational partners that it's human. DeepMind, a Google subsidiary, created an AI that defeated the world champion at the most complicated board game. More recently, AI proved it can be as accurate as trained doctors in diagnosing eye diseases.

And there are any number of stories that warn about a near future where robots will drive all humans into unemployment.

Everywhere you look, AI is conquering new domains, tasks and skills that were previously thought to be the exclusive domain of human intelligence. But does it mean that AI is better than the human mind?

The answer to that question is: It's wrong to compare artificial intelligence to the human mind, because they are totally different things, even if their functions overlap at times.

Artificial Intelligence Is Good at Processing Data, Bad at Thinking in Abstract

Even the most sophisticated AI technology is, at its core, no different from other computer software: bits of data running through circuits at super-fast rates. AI and its popular branch, machine learning and deep learning, can solve any problem as long as you can turn it into the right data sets.

"There's a Huge Difference Between AI and Human Intelligence—so Let's Stop Comparing Them," by Ben Dickson, Tech Talks, August 21, 2018. Reprinted by permission.

Take image recognition. If you give a deep neural network, the structure underlying deep learning algorithms, enough labeled images, it can compare their data in very complicated ways and find correlations and patterns that define each type of object. It then uses that information to label objects in images it hasn't seen before.

The same process happens in voice recognition. Given enough digital samples of a person's voice, a neural network can find the common patterns in the person's voice and determine if future recordings belong to that person.

Everywhere you look, whether it's a computer vision algorithm doing face recognition or diagnosing cancer, an AI-powered cybersecurity tool ferreting out malicious network traffic, or a complicated AI project playing computer games, the same rules apply.

The techniques change and progress: Deep neural networks enable AI algorithms to analyze data through multiple layers; generative adversarial networks (GAN) enable AI to create new data based on the data set it has trained on; reinforcement learning enables AI to develop its own behavior based on the rules that apply to an environment… But what remains the same is the same basic principle: If you can break down a task into data, AI will be able to learn it.

Take note, however, that designing AI models is a complicated task that few people can accomplish. Deep learning engineers and researchers are some of the most coveted and highly paid experts in the tech industry.

Where AI falls short is thinking in the abstract, applying common sense, or transferring knowledge from one area to another. For instance, Google's Duplex might be very good at reserving restaurant tables and setting up appointments with your barber, two narrow and very specific tasks. The AI is even able to mimic natural human behavior, using inflections and intonations as any human speaker would. But as soon as the conversation goes off course, Duplex will be hard-pressed to answer in a coherent way. It will either have to disengage or use the help of a human assistant to continue the conversation in a meaningful way.

There are many proven instances in which AI models fail in spectacular and illogical ways as soon as they're presented with an example that falls outside of their problem domain or is different from the data they've been trained on. The broader the domain, the more data the AI needs to be able to master it, and there will always be edge cases, scenarios that haven't been covered by the training data and will cause the AI to fail.

An example is self-driving cars, which are still struggling to become fully autonomous despite having driven tens of millions of kilometers, much more than a human needs to become an expert driver.

Humans Are Bad at Processing Data, Good at Making Abstract Decisions

Let's start with the data part. Contrary to computers, humans are terrible at storing and processing information. For instance, you must listen to a song several times before you can memorize it. But for a computer, memorizing a song is as simple as pressing "Save" in an application or copying the file into its hard drive. Likewise, unmemorizing is hard for humans. Try as you might, you can't forget bad memories. For a computer, it's as easy as deleting a file.

When it comes to processing data, humans are obviously inferior to AI. In all the examples iterated above, humans might be able to perform the same tasks as computers. However, in the time that it takes for a human to identify and label an image, an AI algorithm can classify one million images. The sheer processing speed of computers enable them to outpace humans at any task that involves mathematical calculations and data processing.

However, humans can make abstract decisions based on instinct, common sense and scarce information. A human child learns to handle objects at a very young age. For an AI algorithm, it takes hundreds of years' worth of training to perform the same task.

For instance, when humans play a video game for the first time in their life, they can quickly transfer their everyday life knowledge into the game's environment, such as staying away from pits, ledges, fire and pointy things (or jumping over them). They know they must dodge bullets and avoid getting hit by vehicles. For AI, every video game is a new, unknown world it must learn from scratch.

Humans can invent new things, including all the technologies that have ushered in the era of artificial intelligence. AI can only take data, compare it, come up with new combinations and presentations, and predict trends based on how previous sequences.

Humans can feel, imagine, dream. They can be selfless or greedy. They can love and hate, they can lie, they forget, they confuse facts. And all of those emotions can change their decisions in rational or irrational ways. They're imperfect and flawed beings made of flesh, which decays with time. But every single one of them is unique in his or her own way and can create things that no one else can.

AI is, at its core, is tiny bursts of electricity running through billions of lifeless circuits.

Let's Stop Comparing AI with Human Intelligence

None of this means that AI is superior to the human brain, or vice versa. They point is, they're totally different things.

AI is good at repetitive tasks that have clearly defined boundaries and can be represented by data, and bad at broad tasks that require intuition and decision-making based on incomplete information.

In contrast, human intelligence is good for settings where you need common sense and abstract decisions, and bad at tasks that require heavy computations and data processing in real time.

Looking at it from a different perspective, we should think about AI as augmented intelligence. AI and human intelligence complement each other, making up for each other's shortcomings. Together, they can perform tasks that none of them could have done individually.

For instance, AI is good at perusing huge amounts of network traffic and pointing out anomalies, but can make mistakes when deciding which ones are the real threats that need investigation. A human analyst, on the other hand, is not very good at monitoring gigabytes of data going through a company's network, but they're adept at relating anomalies to different events and figuring out which ones are the real threats. Together AI and human analysts can fill each other's gaps.

Now, what about all those articles that claim human labor is going instinct? Well, a lot of it is hype, and the facts prove that the expansion of AI is creating more jobs than it is destroying. But it's true that it will obviate the need for humans in many tasks, just as every technological breakthrough has done in the past. But that's probably because those jobs were never meant for humans. We were spending precious human intelligence and labor on those jobs because we hadn't developed the technologies to automate them yet.

As AI becomes adept at performing more and more tasks, we as humans will find more time to put our intelligence to real use, at being creative, being social, at arts, sports, literature, poetry and all the things that are valuable because the human element and character that goes into them. And we'll use our augmented intelligence tools to enhance those creations.

The future will be one where artificial and human intelligence will build together, not apart.

AI Will Be Human-Like, but Better

Ana Santos Rutschman

Ana Santos Rutschman is Jaharis Faculty Fellow in Health Law and Intellectual Property at DePaul University. She researches and writes on topics related to emerging health technologies, with a particular focus on vaccines and other forms of biotechnology.

The late Stephen Hawking was a major voice in the debate about how humanity can benefit from artificial intelligence. Hawking made no secret of his fears that thinking machines could one day take charge. He went as far as predicting that future developments in AI "could spell the end of the human race."

But Hawking's relationship with AI was far more complex than this often-cited soundbite. The deep concerns he expressed were about superhuman AI, the point at which AI systems not only replicate human intelligence processes, but also keep expanding them, without our support—a stage that is at best decades away, if it ever happens at all. And yet Hawking's very ability to communicate those fears, and all his other ideas, came to depend on basic AI technology.

Hawking's Conflicted Relationship with AI

At the intellectual property and health law centers at DePaul University, my colleagues and I study the effects of emerging technologies like the ones Stephen Hawking worried about. At its core, the concept of AI involves computational technology designed to make machines function with foresight that mimics, and ultimately surpasses, human thinking processes.

Hawking cautioned against an extreme form of AI, in which thinking machines would "take off" on their own, modifying

"Stephen Hawking Warned About the Perils of Artificial Intelligence—Yet AI Gave Him a Voice," by Ana Santos Rutschman, The Conversation, March 15, 2018. https://theconversation.com/stephen-hawking-warned-about-the-perils-of-artificial-intelligence-yet-ai-gave-him-a-voice-93416. Licensed under CC BY-ND-4.0 International.

themselves and independently designing and building ever more capable systems. Humans, bound by the slow pace of biological evolution, would be tragically outwitted.

AI as a Threat to Humanity?

Well before it gets to the point of superhuman technology, AI can be put to terrible uses. Already, scholars and commentators worry that self-flying drones may be precursors to lethal autonomous robots.

Today's early stage AI raises several other ethical and practical problems, too. AI systems are largely based on opaque algorithms that make decisions even their own designers may be unable to explain. The underlying mathematical models can be biased, and computational errors may occur. AI may progressively displace human skills and increase unemployment. And limited access to AI might increase global inequality.

The One Hundred Year Study on Artificial Intelligence, launched by Stanford University in 2014, highlighted some of these concerns. But so far it has identified no evidence that AI will pose any "imminent threat" to humankind, as Hawking feared.

Still, Hawking's views on AI are somewhat less alarmist and more nuanced than he usually gets credit for. At their heart, they describe the need to understand and regulate emerging technologies. He repeatedly called for more research on the benefits and dangers of AI. And he believed that even non-superhuman AI systems could help eradicate war, poverty and disease.

Hawking Talks

This apparent contradiction—a fear of humanity being eventually overtaken by AI but optimism about its benefits in the meantime—may have come from his own life: Hawking had come to rely on AI to interact with the world.

Unable to speak since 1985, he used a series of different communication systems that helped him talk and write, culminating in the now-legendary computer operated by one muscle in his right cheek.

The first iteration of the computer program was exasperatingly slow and prone to errors. Very basic AI changed that. An open-source program made his word selection significantly faster. More importantly, it used artificial intelligence to analyze Hawking's own words, and then used that information to help him express new ideas. By processing Hawking's books, articles and lecture scripts, the system got so good that he did not even have to type the term people most associate with him, "the black hole." When he selected "the," "black" would automatically be suggested to follow it, and "black" would prompt "hole" onto the screen.

AI Improves People's Health

Stephen Hawking's experience with such a basic form of AI illustrates how non-superhuman AI can indeed change people's lives for the better. Speech prediction helped him cope with a devastating neurological disease. Other AI-based systems are already helping prevent, fight and lessen the burden of disease.

For instance, AI can analyze medical sensors and other health data to predict how likely a patient is to develop a severe blood infection. In studies it was substantially more accurate—and provided much more advance warning—than other methods.

Another group of researchers created an AI program to sift through electronic health records of 700,000 patients. The program, called "Deep Patient," unearthed linkages that had not been apparent to doctors, identifying new risk patterns for certain cancers, diabetes and psychiatric disorders.

AI has even powered a robotic surgery system that outperformed human surgeons in a procedure on pigs that's very similar to one type of operation on human patients.

There's so much promise for AI to improve people's health that collecting medical data has become a cornerstone of both software development and public-health policy in the U.S. For example, the Obama White House launched a research effort seeking to collect DNA from at least a million Americans. The data will be made available for AI systems to analyze when studying new medical treatments, potentially improving both diagnoses and patients' recovery.

All of these benefits from AI are available right now, and more are in the works. They do suggest that superhuman AI systems could be extremely powerful, but despite warnings from Hawking and fellow technology visionary Elon Musk that day may never come. In the meantime, as Hawking knew, there is much to be gained. AI gave him a better and more efficient voice than his body was able to provide, with which he called for both research and restraint.

CHAPTER 3

Will Artificial Intelligence Change the World of Work?

Overview: AI's Influence on the Labor Market Will Require Changes to Education

Elizabeth Mann Levesque

Elizabeth Mann Levesque is a student support and classroom climate consultant at the University of Michigan. She previously served as a nonresident fellow at the Brookings Institution.

The growth of artificial intelligence (AI) and emerging technologies (ET) is poised to reshape the workforce.[1] While the exact impact of AI and ET is unclear, experts expect that many jobs currently performed by humans will be performed by robots in the near future, and at the same time, new jobs will be created as technology advances. These impending changes have important implications for the field of education. Schools must prepare students to remain competitive in the labor market, and postsecondary institutions must provide students and displaced workers with relevant education and retraining opportunities. Innovations in technology will also create new tools to support educators, students, and others seeking retraining and employment.

Consequently, there is a multitude of policy-relevant questions that we may consider with respect to how AI and ET will impact education. Rather than focus on just one of these many questions, this paper provides an overview of some of the most salient issues we should consider with respect to what technological advances in AI and ET mean for education. Specifically, this paper discusses several types of challenges, opportunities, and risks that AI and ET pose to the field of education. This paper then concludes with several recommendations for adapting education in anticipation of the changes associated with advances in AI and ET.

"The Role of AI in Education and the Changing US Workforce," by Elizabeth Mann Levesque, The Brookings Institution, October 18, 2018. Reprinted by permission.

Central Challenges Facing Education

The types of jobs that are at the least risk of being replaced by automation involve problem solving, teamwork, critical thinking, communication, and creativity.[2] The education profession is unlikely to see a dramatic drop in demand for employees given the nature of work in this field. Rather, preparing students for the changing labor market will likely be a central challenge for schools and educators. Policymakers and practitioners must adapt K-12 education to help students develop the skills that are likely to remain in demand (sometimes referred to as "21st century skills"). K-12 education should thus prioritize teaching critical thinking, problem solving, and teamwork across subject areas. Teaching students to become analytical thinkers, problem solvers, and good team members will allow them to remain competitive in the job market even as the nature of work changes. Equally important, these skills form a strong foundation for independent thinking that will serve students well no matter what career(s) they pursue throughout their lives.

In addition, an increasing demand for technologically skilled workers likely means that proficiency in education in science, technology, engineering, and mathematics (STEM) subjects can position students to be competitive in the workforce. Education in STEM subjects should certainly be a priority, particularly given low levels of proficiency nationwide and large achievement gaps. However, given the increasing importance of developing critical thinking skills that span multiple subject areas, providing high-quality instruction in the STEM fields is only part of the solution to preparing students for the changing workforce.

Further, providing Americans with opportunities for lifelong learning will be central to helping displaced workers find new career pathways. Darrell West explains this shifting employment terrain: "In the contemporary world, people can expect to switch jobs, see whole sectors disrupted, and need to develop additional skills as a result of economic shifts. The type of work they do at age 30 likely will be substantially different from what they do at ages 40, 50, or 60."[3] Individuals need to be equipped to navigate this ever-changing environment, which requires them to identify alternative careers, enroll in and complete relevant education and training programs, and find jobs upon graduation.

Completing these steps is easier said than done. Anthony Carnevale, Director of the Georgetown University Center on Education and the Workforce, explains: "Educational pathways are largely disconnected from the job market, which inhibits students' ability to see their future career pathways lucidly. Policymakers, postsecondary officials and students are not provided with data that keeps them informed."[4] As new technologies continue to reshape the nature of work and the types of jobs available for humans, it is increasingly important to design policies and programs that help individuals find and complete appropriate education, career, and retraining pathways.

Opportunities Through AI and ET

While changes in AI and ET create challenges, these technological advancements also provide opportunities. First, innovations in artificial intelligence can provide teachers with valuable resources. Blended learning, defined as "the strategic integration of in-person learning with technology to enable real-time data use, personalized instruction, and mastery-based progression," uses emerging technology to help teachers personalize education for individual students. This approach is generally known as personalized learning. Studies have found that personalized learning is a promising approach, although implementation challenges remain. Rigorous evaluation of ongoing experimentation with blended and personalized learning will be critical to developing effective approaches to using technology in the classroom. One lesson, described in multiple analyses, is the importance of supporting teachers and educators in using technology to enhance their instruction.

Second, AI and other emerging technologies can be used to create scalable resources that support large numbers of students and others as they navigate education, training, and career pathways. Promising innovations include a conversational AI system that uses personalized text message outreach to help incoming college students complete required pre-matriculation tasks. The University of Virginia, through its "nudge4" center, is working on innovations that leverage technologies to support students and others, such as a current project that seeks to use machine learning to provide personalized transfer guidance to community college students. Interactive online resources, such as the Skillful Initiative, provide resources for job seekers, employers, and career coaches. These innovations and others like them suggest that even as the workforce evolves due to changes in technology, policymakers and educators can simultaneously leverage technological advancements to support students on their higher education pathways and to connect adults with education and career opportunities.

Risks Associated with AI and ET

While AI and ET allow for innovations in supporting students and job seekers, impending changes in the workforce also pose substantial risks. First, existing educational inequities may worsen, accompanied by negative downstream consequences. Nationally representative assessments reveal large and persistent gaps in student achievement by race and income. High-income students are also more likely than low-income students to complete college, even as completion rates among low-income students have risen.[5] Compounding this problem, differences in school resources that correlate with residential and income segregation mean it is likely that the schools best poised to prepare students for changes in the workforce are those that serve children from higher income families.

In this context, policies designed to help students prepare for the future workforce that fail to account for existing inequalities will likely perpetuate these inequalities. For example, based on research about how students responded to newly available information on earnings via the College Scorecard, researchers Harry Holzer and Sandy Baum argue that "[j]ust making general information available is unlikely to significantly improve the college decisions of students from less-privileged backgrounds."[6] Without sufficient attention from policymakers, existing inequalities may widen as the job market contracts.

Second, adapting education to meet changes in the workforce carries the risk of creating overly narrow education goals. In the United States, preparing students to enter the workforce has long been one of the central goals of education. But it is not the only goal, nor should it be. Additional goals include preparing students to engage productively with other members of society and to participate in civic life and the democratic process. For now, there seems to be at least some bipartisan consensus that the goals of education include but are broader than workforce development, reflected in the lack of movement on (and criticism of) the Trump administration's recent proposal to merge the Departments of Education and Labor. Anticipated changes in the workforce wrought by advances in technology do not require us to abandon longstanding goals of education.[7]

Recommendations

Given these challenges, risks, and opportunities, this paper makes several recommendations with respect to education policy to help students and workers adapt to changes in the workforce given advances in AI and ET.

Recommendation 1

State standards and curricula should incorporate 21st century skills across subject areas. The Every Student Succeeds Act, passed in 2015, increased states' flexibility in determining how to hold schools accountable for student learning. While high-stakes tests (particularly in math and reading) are still likely to inform what students learn, schools have renewed latitude to focus on 21st century skills such as critical thinking, problem solving, communication, and teamwork. States can prioritize these skills by incorporating them into subject-area standards and curricula. Multiple resources exist to facilitate these changes in different subject areas, including the Next Generation Science Standards and the College, Career, and Civic Life Framework for Social Studies State Standards. This is not a prescription for states to adopt a certain set of standards. Rather, these standards and frameworks may serve as valuable resources for states seeking to help students develop the skills that will likely be in high demand as AI and ET reshape the workforce.

Under ESSA, schools have renewed latitude to focus on 21st century skills such as critical thinking, problem solving, communication, and teamwork. States can prioritize these skills by incorporating them into subject-area standards and curricula.

Recommendation 2
Federal legislation and policy should explore and support workforce development partnerships. Building partnerships between educators at the postsecondary level and employers is crucial for providing students with opportunities to pursue careers that are likely to remain available to humans. At the same time, creating meaningful partnerships requires investments in time and resources on both sides.[8] Building a strong evidence base about how to design and implement effective partnerships between employers and two- and four-year colleges can help convince schools and employers to invest in these partnerships and can support them in providing effective programs.

The federal government can provide valuable leadership in this area. The Trump administration's National Council on the American Worker could explore promising models of these partnerships, convene educators and employers to discuss potential paths forward, and identify concrete policies that the federal government could pursue to facilitate innovation in this area. The Institute of Education Sciences (IES), the research arm of the Department of Education, recently announced plans to expand coverage of postsecondary education "especially with regard to career and technical training" in the What Works Clearinghouse, a federal repository of evidence-based research on education. This expanded coverage could include research on partnerships between postsecondary institutions and employers, with the aim of understanding how to create a system in which students successfully navigate from college to careers.

Recommendation 3
Support displaced workers and other "non-traditional students" in their search for new career pathways. As advances in AI displace current members of the workforce, people who have been in the workforce for years will need support in finding new careers. Given the difficulty of navigating education and career pathways, supporting these individuals should be a high priority. Existing resources, such as one-stop career centers provided for under the Workforce Innovation and Opportunities Act (WIOA), will become increasingly important. Continued investment in these brick and mortar resources alongside rigorous evaluation of their programming will improve our knowledge of how to design and provide effective career services. WIOA also allocates funds to study the effectiveness of workforce development systems with respect to "assisting workers to obtain the skills needed to utilize emerging technologies." This funding is one example of how federal policy can support research and development of programs designed to support workers as the type of available work changes.

Creating and disseminating online resources designed specifically for this population is also important. For example, Georgetown's Carnevale advocates creating online matching systems that "tie job exchanges (online job-search engines) to learning exchanges that match job openings and career pathways to available courses offered by postsecondary institutions in the classroom and online." Other online resources such as the Skillful Initiative, mentioned above, provide support for job seekers, employers, and career coaches. Building an evidence base through rigorous evaluation of these and similar programs will be crucial to identifying effective support systems.

Conclusion

This paper has discussed several of the major challenges and opportunities that AI and ET pose to educators at the K-12 and postsecondary levels, given predicted changes in the workforce. Two broad lessons arise from this discussion. First, leveraging advances in AI and ET may allow for scalable programs capable of reaching many different types of individuals who will need support adapting to changes in the workforce. Second, investing in innovation and evaluation of promising programs to address these challenges should be a high priority for the philanthropic and business communities as well as for policymakers at every level of government.

Footnotes

1. According to Darrell West and John Allen, who draw on Shubhendu and Vijay, AI generally refers to "machines that respond to stimulation consistent with traditional responses from humans, given the human capacity for contemplation, judgment and intention." Further, these systems "make decisions which normally require human level of expertise" and "operate in an intentional, intelligent, and adaptive manner."

2. According to a 2017 analysis by the McKinsey Global Institute, the types of tasks that have the least potential for automation include managing and developing people, applying expertise to decision making, planning, and creative tasks, interfacing with stakeholders (such as greeting customers at a store or explaining service information to customers from a call center), and performing physical tasks or operating machinery in unpredictable physical environments (p. 42).

3. West 2018, p. 109.

4. In a recent report, the Council of Economic Advisers similarly identifies an information gap as a barrier in connecting workers to promising jobs and careers.

5. Bailey and Dynarski 2011.

6. Holzer and Baum 2017, p. 123.

7. Conversations about preparing students for the workforce will likely engage with other questions about what the goals of education are, such as whether attending a four-year college should be the ultimate goal. These are difficult questions without clear answers; not every good-paying career requires a four-year degree, but all students deserve a K-12 education that prepares them to enroll in and complete a four-year degree program. Addressing these questions in discussions over how to prepare the future workforce can help avoid negative unintended consequences, such as tracking some students into vocational pathways rather than giving all students the opportunity to prepare for a four-year degree program.

8. Wyner 2014.

A Significant Loss of Jobs Will Occur in the Future

Calum McClelland

Calum McClelland is head of operations at IoT for All. He is deeply interested in the moral ramifications of new technologies and believes in leveraging the Internet of Things to help build a better world for everyone.

Artificial intelligence (AI) is no longer a thing of science fiction. It exists in the world all around us, automating simple tasks and dramatically improving our lives. But as AI and automation becomes increasingly capable, how will this alternative labor source affect your future workforce? In this article, we'll take a look at both some optimistic and pessimistic views of the future of our jobs amidst increasing AI capabilities.

Technology-driven societal changes, like what we're experiencing with AI and automation, always engender concern and fear—and for good reason. A two-year study from McKinsey Global Institute suggests that by 2030, intelligent agents and robots could replace as much as 30 percent of the world's current human labor. McKinsey suggests that, in terms of scale, the automation revolution could rival the move away from agricultural labor during the 1900s in the United States and Europe, and more recently, the explosion of the Chinese labor economy.

McKinsey reckons that, depending upon various adoption scenarios, automation will displace between 400 and 800 million jobs by 2030, requiring as many as 375 million people to switch job categories entirely. How could such a shift not cause fear and concern, especially for the world's vulnerable countries and populations?

"The Impact of Artificial Intelligence—Widespread Job Losses," by Calum McClelland, IoT for All, July 1, 2020. Reprinted by permission.

The Brookings Institution suggests that even if automation only reaches the 38 percent means of most forecasts, some Western democracies are likely to resort to authoritarian policies to stave off civil chaos, much like they did during the Great Depression. Brookings writes, "The United States would look like Syria or Iraq, with armed bands of young men with few employment prospects other than war, violence, or theft." With frightening yet authoritative predictions like those, it's no wonder AI and automation keeps many of us up at night.

"Stop Being a Luddite"

The Luddites were textiles workers who protested against automation, eventually attacking and burning factories because "they feared that unskilled machine operators were robbing them of their livelihood." The Luddite movement occurred all the way back in 1811, so concerns about job losses or job displacements due to automation are far from new.

When fear or concern is raised about the potential impact of artificial intelligence and automation on our workforce, a typical response is thus to point to the past; the same concerns are raised time and again and prove unfounded.

In 1961, President Kennedy said, "the major challenge of the sixties is to maintain full employment at a time when automation is replacing men." In the 1980s, the advent of personal computers spurred "computerphobia" with many fearing computers would replace them.

So what happened?

Despite these fears and concerns, every technological shift has ended up creating more jobs than were destroyed. When particular tasks are automated, becoming cheaper and faster, you need more human workers to do the other functions in the process that haven't been automated.

> During the Industrial Revolution more and more tasks in the weaving process were automated, prompting workers to focus on the things machines could not do, such as operating

a machine, and then tending multiple machines to keep them running smoothly. This caused output to grow explosively. In America during the 19th century the amount of coarse cloth a single weaver could produce in an hour increased by a factor of 50, and the amount of labour required per yard of cloth fell by 98%. This made cloth cheaper and increased demand for it, which in turn created more jobs for weavers: their numbers quadrupled between 1830 and 1900. In other words, technology gradually changed the nature of the weaver's job, and the skills required to do it, rather than replacing it altogether.
—*The Economist*, Automation and Anxiety

Impact of Artificial Intelligence—A Bright Future?

Looking back on history, it seems reasonable to conclude that fears and concerns regarding AI and automation are understandable but ultimately unwarranted. Technological change may eliminate specific jobs, but it has always created more in the process.

Beyond net job creation, there are other reasons to be optimistic about the impact of artificial intelligence and automation.

Simply put, jobs that robots can replace are not good jobs in the first place. As humans, we climb up the rungs of drudgery—physically tasking or mind-numbing jobs—to jobs that use what got us to the top of the food chain, our brains.
—*The Wall Street Journal*, The Robots Are Coming. Welcome Them.

By eliminating the tedium, AI and automation can free us to pursue careers that give us a greater sense of meaning and well-being. Careers that challenge us, instill a sense of progress, provide us with autonomy, and make us feel like we belong; all research-backed attributes of a satisfying job.

And at a higher level, AI and automation will also help to eliminate disease and world poverty. Already, AI is driving great advances in medicine and healthcare with better disease prevention, higher accuracy diagnosis, and more effective treatment and cures. When it comes to eliminating world poverty, one of the biggest barriers is identifying where help is needed most. By applying AI analysis to data from satellite images, this barrier can be surmounted, focusing aid most effectively.

Impact of Artificial Intelligence—A Dark Future

I am all for optimism. But as much as I'd like to believe all of the above, this bright outlook on the future relies on seemingly shaky premises. Namely:

1. The past is an accurate predictor of the future.
2. We can weather the painful transition.
3. There are some jobs that only humans can do.

The Past Isn't an Accurate Predictor of the Future

As explored earlier, a common response to fears and concerns over the impact of artificial intelligence and automation is to point to the past. However, this approach only works if the future behaves similarly. There are many things that are different now than in the past, and these factors give us good reason to believe that the future will play out differently.

In the past, technological disruption of one industry didn't necessarily mean the disruption of another. Let's take car manufacturing as an example; a robot in automobile manufacturing can drive big gains in productivity and efficiency, but that same robot would be useless trying to manufacture anything other than a car. The underlying technology of the robot might be adapted, but at best that still only addresses manufacturing

AI is different because it can be applied to virtually any industry. When you develop AI that can understand language, recognize patterns, and problem solve, disruption isn't contained. Imagine creating an AI that can diagnose disease and handle medications, address lawsuits, and write articles like this one. No need to imagine: AI is already doing those exact things.

Another important distinction between now and the past is the speed of technological progress. Technological progress doesn't advance linearly, it advances exponentially. Consider Moore's Law: the number of transistors on an integrated circuit doubles roughly every two years.

In the words of University of Colorado physics professor Albert Allen Bartlett, "The greatest shortcoming of the human race is our inability to understand the exponential function." We drastically underestimate what happens when a value keeps doubling.

What do you get when technological progress is accelerating and AI can do jobs across a range of industries? An accelerating pace of job destruction.

> There's no economic law that says "You will always create enough jobs or the balance will always be even," it's possible for a technology to dramatically favour one group and to hurt another group, and the net of that might be that you have fewer jobs.
> —Erik Brynjolfsson, Director of the MIT Initiative on the Digital Economy

In the past, yes, more jobs were created than were destroyed by technology. Workers were able to reskill and move laterally into other industries instead. But the past isn't always an accurate predictor of the future. We can't complacently sit back and think that everything is going to be ok.

Which brings us to another critical issue …

The Transition Will Be Extremely Painful

Let's pretend for a second that the past actually will be a good predictor of the future; jobs will be eliminated but more jobs will be created to replace them. This brings up an absolutely critical question, what kinds of jobs are being created and what kinds of jobs are being destroyed?

> Low- and high-skilled jobs have so far been less vulnerable to automation. The low-skilled jobs categories that are considered to have the best prospects over the next decade—including food service, janitorial work, gardening, home health, childcare, and security—are generally physical jobs, and require face-to-face interaction. At some point robots will be able to fulfill these

roles, but there's little incentive to roboticize these tasks at the moment, as there's a large supply of humans who are willing to do them for low wages.

—Slate, Will Robots Steal Your Job?

Blue-collar and white-collar jobs will be eliminated—basically, anything that requires middle-skills (meaning that it requires some training, but not much). This leaves low-skill jobs, as described above, and high-skill jobs that require high levels of training and education.

There will assuredly be an increasing number of jobs related to programming, robotics, engineering, etc. After all, these skills will be needed to improve and maintain the AI and automation being used around us.

But will the people who lost their middle-skilled jobs be able to move into these high-skill roles instead? Certainly not without significant training and education. What about moving into low-skill jobs? Well, the number of these jobs is unlikely to increase, particularly because the middle-class loses jobs and stops spending money on food service, gardening, home health, etc.

The transition could be very painful. It's no secret that rising unemployment has a negative impact on society; less volunteerism, higher crime, and drug abuse are all correlated. A period of high unemployment, in which tens of millions of people are incapable of getting a job because they simply don't have the necessary skills, will be our reality if we don't adequately prepare.

So how do we prepare? At the minimum, by overhauling our entire education system and providing means for people to re-skill.

To transition from 90% of the American population farming to just 2% during the first industrial revolution, it took the mass introduction of primary education to equip people with the necessary skills to work. The problem is that we're still using an education system that is geared for the industrial age. The three Rs (reading, writing, arithmetic) were once the important skills to learn to succeed in the workforce. Now, those are the skills quickly being overtaken by AI.

In addition to transforming our whole education system, we should also accept that learning doesn't end with formal schooling. The exponential acceleration of digital transformation means that learning must be a lifelong pursuit, constantly re-skilling to meet an ever-changing world.

Making huge changes to our education system, providing means for people to re-skill, and encouraging lifelong learning can help mitigate the pain of the transition, but is that enough?

Will All Jobs Be Eliminated?

When I originally wrote this article a couple of years ago, I believed firmly that 99% of all jobs would be eliminated. Now, I'm not so sure. Here was my argument at the time:

> The claim that 99% of all jobs will be eliminated may seem bold, and yet it's all but certain. All you need are two premises:
>
> 1. We will continue making progress in building more intelligent machines.
> 2. Human intelligence arises from physical processes.

The first premise shouldn't be at all controversial. The only reason to think that we would permanently stop progress, of any kind, is some extinction-level event that wipes out humanity, in which case this debate is irrelevant. Excluding such a disaster, technological progress will continue on an exponential curve. And it doesn't matter how fast that progress is; all that matters is that it will continue. The incentives for people, companies, and governments are too great to think otherwise.

The second premise will be controversial, but notice that I said human intelligence. I didn't say "consciousness" or "what it means to be human." That human intelligence arises from physical processes seems easy to demonstrate: if we affect the physical processes of the brain we can observe clear changes in intelligence. Though a gloomy example, it's clear that poking holes in a person's brain results in changes to their intelligence. A well-placed poke in someone's Broca's area and voilà—that person can't process speech.

With these two premises in hand, we can conclude the following: we will build machines that have human-level intelligence and higher. It's inevitable.

We already know that machines are better than humans at physical tasks, they can move faster, more precisely, and lift greater loads. When these machines are also as intelligent as us, there will be almost nothing they can't do—or can't learn to do quickly. Therefore, 99% of jobs will eventually be eliminated.

But that doesn't mean we'll be redundant. We'll still need leaders (unless we give ourselves over to robot overlords) and our arts, music, etc., may remain solely human pursuits too. As for just about everything else? Machines will do it—and do it better.

"But who's going to maintain the machines?" The machines.

"But who's going to improve the machines?" The machines.

Assuming they could eventually learn 99% of what we do, surely they'll be capable of maintaining and improving themselves more precisely and efficiently than we ever could.

The above argument is sound, but the conclusion that 99% of all jobs will be eliminated I believe over-focused on our current conception of a "job." As I pointed out above, there's no guarantee that the future will play out like the past. After continuing to reflect and learn over the past few years, I now think there's good reason to believe that while 99% of all current jobs might be eliminated, there will still be plenty for humans to do (which is really what we care about, isn't it?).

> The one thing that humans can do that robots can't (at least for a long while) is to decide what it is that humans want to do. This is not a trivial semantic trick; our desires are inspired by our previous inventions, making this a circular question.
> —*The Inevitable: Understanding the 12 Technological Forces That Will Shape Our Future,* by Kevin Kelly

Perhaps another way of looking at the above quote is this: a few years ago I read the book *Emotional Intelligence*, and was shocked to discover just how essential emotions are to decision making. Not just important, essential. People who had experienced brain damage to the emotional centers of their brains were absolutely incapable of making even the smallest decisions. This is because, when faced with a number of choices, they could think of logical reasons for doing or not doing any of them but had no emotional push/pull to choose.

So while AI and automation may eliminate the need for humans to do any of the doing, we will still need humans to determine what to do. And because everything that we do and everything that we build sparks new desires and shows us new possibilities, this "job" will never be eliminated.

If you had predicted in the early 19th century that almost all jobs would be eliminated, and you defined jobs as agricultural work, you would have been right. In the same way, I believe that what we think of as jobs today will almost certainly be eliminated too. But this does not mean that there will be no jobs at all, the "job" will instead shift to determining, what do we want to do? And then working with our AI and machines to make our desires a reality.

Is this overly optimistic? I don't think so. Either way, there's no question that the impact of artificial intelligence will be great and it's critical that we invest in the education and infrastructure needed to support people as many current jobs are eliminated and we transition to this new future.

Work Will Be Transformed by Artificial Intelligence

Dennis Spaeth

Dennis Spaeth is the owner and publisher of Cutting Tool Engineering *magazine.*

It may seem like an inevitable fact of administrative positions that anyone who fills them will be subjected to a never-ending litany of repetitive tasks. Employees in these jobs often don't receive work that engages their brains or peaks their interests. Rather than flexing their critical thinking skills, these workers are resigned to completing the necessary, yet boring, administrative tasks.

However, this rather bleak outlook is on the verge of changing with the advent of artificial intelligence. While the world's first instance of robotics can be traced back to the invention of the Egyptian water clock over 3,500 years ago in 1500 BCE, AI wasn't coined as a term until 1956. In this more modern robotic iteration, the idea was that a computer system can learn from each experience it encounters once it has a basis built from human ingenuity.

AI surrounds us today more than ever before. Virtual assistants, such as Siri and Alexa, have become commonplace devices that help homeowners do everything from creating a shopping list to turning up the thermostat. In the workplace, artificial intelligence has the potential to transform workforce productivity. However, as with any technological advancement, AI also has its pitfalls.

Before you take the side of being pro- or anti-AI, take a look at some of the advantages and disadvantages of the integration of artificial intelligence into the modern workforce.

"Artificial Intelligence Is Transforming the Workforce as We Know It," by Dennis Spaeth, Workplace Insight, March 18, 2019. Reprinted by permission.

An Increase in Productivity

One of the most noteworthy benefits of AI is the increase in productivity it provides businesses. A human can effectively perform up to about 10 hours of repetitive work on a good day. This number is even less on average when you factor in labor laws and the days when a person is feeling more tired than usual. A system powered by artificial intelligence can work for an indefinite amount of time. Robots don't need rest and are not governed by regulations that limit the number of hours they can work in a week.

This ability to keep working at all hours of the day translates into a major increase in productivity for businesses. More of those menial tasks will be completed as AI works through the day and night with few interruptions. The only obstacle a business may face is when the machine's code becomes corrupted and needs to be reset. However, the amount of time it takes to turn the system off and back on again is significantly less than the hours a human spends away from work.

AI also increases productivity by freeing up humans to take on more tasks that involve critical thinking. When workers don't have to fill their time with the repetitive tasks on which a business relies, they can instead use their ingenuity and creativity to solve problems and contribute to innovative efforts. Not only is this work more engaging for the workers, but it is more valuable for the company in the end.

Error-Free Processing

To err is human. The same does not apply to computers. While humans may make an error in data processing due to fatigue, lack of knowledge, or any other factor, AI only makes a mistake if the person who programmed it made it first. Even this possibility for error is slight. The technology that makes AI possible is built on a language of mathematics and commands. This language doesn't leave much room for error and the automated nature of the system greatly diminishes the chances for mistakes when compared to a human.

Although it may not seem like a human error would affect a business all that much, even the smallest one can have a negative ripple effect. One mistake in processing can skew a business's numbers dramatically. Even the less extreme errors tend to cost a business money they could have used elsewhere. For any company looking to reduce costs associated with human error, turning to AI may be the solution.

Decreased Number of Jobs

Artificial intelligence has the potential to create more jobs for humans by allowing them to focus on more creative endeavors, but it is certain that the implementation of AI will also result in a significant number of job losses. According to Gartner, a global research and advisory firm, AI could create over 500,000 jobs in the United States. However, other analysts estimate that net job losses could number over 1 million if AI takes over parts of the workforce. Not only will repetitive administrative positions go to these advanced computer systems, but jobs that require less skill in places such as factories will also shift to the robots.

This shift will disproportionately affect those who hold more low-skill jobs. As people from low-income areas who did not pursue higher education tend to hold more of these jobs, they will be the ones most heavily impacted by the incorporation of AI into the workforce. For instance, driverless cars are already making their way onto the roads. Without the need for full-time human drivers, immigrants who tend to hold many of these positions will be without work as computer systems replace them.

Inability to Make Judgement Calls

Artificial intelligence is capable of many things, but it may never be capable of making fair judgement calls in the way that humans can. While the human brain weighs the pros and cons of a situation and can make a decision based on the the circumstances at hand, AI will make a decision as it is programmed to do. Even the best computer system won't be able to consider the minute, and often human, factors that impact a decision. The computer will simply use the equations with which a human initially programmed it to come to its conclusion.

An example of a bad AI judgement called occurred in Australia in 2014. There was a shooting in downtown Sydney and many people started to flee the area by ordering rides through Uber. As the demand in the concentrated area surged, Uber's algorithms applied its trusted supply-and-demand surge charges and ride rates skyrocketed. The calculations couldn't factor in the emergency situation happening downtown and the system charged victims outrageous prices for trying to leave a dangerous environment.

While humans can make adjustments to the algorithms now that they know this situation is a possibility, it doesn't make it better for the people who had to go through that traumatic experience. Unless all solutions are anticipated and programmed from the start, AI can work to the detriment of humans or the surrounding environment.

There is no doubt that technology has worked throughout human history to advance society and improve quality of life. Artificial intelligence has the potential to do the same and has already done so for many people through the use of virtual assistants and other AI applications. However, the downsides to the technology are not something to ignore. The only thing that is certain is that AI can change the workforce for good, whether humans are ready for it or not.

Automation Doesn't Necessarily Mean Massive Job Loss

Carlos Bonilla

Carlos Bonilla is an analyst at Econsult Solutions. He specializes in economic and environmental issues of urban areas.

Recent advancements in robotics and artificial intelligence have some experts worrying about the coming obsolescence of the human worker. Some are even calling for a universal basic income provided by the government for everyone, under the assumption that work will become scarce. But economists seem to have a pretty different view from the futurist and Silicon Valley types.

One of the reasons that economists are skeptical is because much of the fear that has recently been stoked in the media is familiar. Every few decades, predictions about the "end of work" have pervaded public discourse, and they've always been wrong. There was a spike of automation anxiety in the late 1920s and '30s, when machines were starting to take over jobs on farms and in factories. Automation anxiety surged again in the late '50s and '60s, when President Kennedy ranked automation as the major domestic challenge of the time.

We find ourselves in another era of automation anxiety right now. Is there still reason to be skeptical or is this time any different?

Jobs Lost to Automation

Ever since women joined the workforce in big numbers, the share of those working has stayed around 80% (outside of recessions). During this period in the US, technology displaced millions of farmers, telephone operators, and factory and railroad workers. We

---

"Will Automation Lead to Mass Unemployment?" Econsult Solutions, Inc., February 14, 2020. Reprinted by permission.

have also since lost pinsetters and elevator operators. But despite the number of jobs lost, work persists.

It's worth noting that technology and automation have consistently transformed the way work gets done. Robots and automation have allowed us to increase efficiency by making more things for less. But are machines replacing human labor at a faster pace than the economy can absorb? Current evidence indicates that while in the short term technology often does displace jobs, it also creates many new jobs long-term.

Jobs Created by Automation

New technology can create jobs in a few ways. There are the direct jobs for people who design and maintain the technology, and sometimes whole new industries built on the technology. But the part that is often overlooked is the indirect effect of labor saving inventions. When companies can do more with less, they can expand (i.e., add new products or open new locations) and lower prices to compete. And when goods and services are cheaper, consumers can afford to buy more of their product, or use their savings to spend on other things (like sporting events, dining, etc.).

ATMs are an example of this in practice. When ATM machines were introduced in the 1980s, it was thought that bank tellers would quickly stop existing. Over time, each bank branch did end up employing fewer tellers. However, ATMs made it cheaper to operate bank branches, which caused more bank branches to open. This unexpectedly lead to more bank teller jobs overall. Similarly, when spreadsheet software was introduced, it displaced two million bookkeepers. However, it also created millions of new jobs in the form of accountants, auditors and financial analysts.

Warnings about the "end of work" tend to focus on the disruptiveness of the new technology itself and not the new productivity brought on by it. The widely known Oxford study from 2013, for instance, is often quoted stating that nearly half of all jobs are vulnerable to automation. But even that study assesses only the capabilities of automation technology; it does not attempt to estimate the actual extent or pace of automation or the overall effect on employment. In other words, it only looks at the kinds of tasks ripe for automation, not the actual jobs (which contain a variety of different kinds of tasks, many of which are still hard to automate) themselves.

The key argument here is that automation displaces workers who are doing highly automatable work and tasks, but it does not affect the *total* number of jobs in the economy because of offsetting effects. This process is how our standard of living has improved over time, and it has always required workers. It is important to keep in mind that even though technology can be a net job creator, it does not mean that the new jobs created will show up right away, be located in the same place or even pay the same as the ones that were lost. All it means is that the overall need for human work has not gone away.

Accelerating Change

As noted above, economists tend to be skeptical of fears related to technological unemployment, pointing to data that suggests this particular scenario is nowhere in sight. While technologists and futurists acknowledge the validity of these conclusions, they do question whether historical trends are a good guide of what's to come. After all, the rate of technological change and its effect on the economy has been grossly underestimated before.

In the mid-2000s, two economists assessed the future of automation and concluded that tasks like driving in traffic would be "enormously difficult" to teach to a computer. Around the same time, a review of fifty years of research concluded that human level speech recognition "has proved to be an elusive goal." Today, we know these predictions were wrong given recent advances in artificial intelligence. Since the start of the 2010s, artificial intelligence has beaten professional gamers, sparked a new age of digital innovation, and is now powering the self-driving car movement. And its footprint can be felt in a number of industries, including the Internet of Things (IoT), transportation, logistics, health, and fintech.

This technology is behind much of the automation happening today, and it has the potential to be a key disruptive force in the near future. This is because, unlike the automation that occurred during the Industrial Revolution (which complemented or replaced physical work), modern automation is substituting cognitive work. While advances in artificial intelligence are certainly happening, the real modern advances aren't in jobs involving manual labor. Rather, it is in jobs like accounting, bookkeeping, and other related roles. With machines increasingly mirroring human cognition, some argue that this will have many (potentially negative) consequences for our service and knowledge based economy.

[…] [T]he number of transistors that engineers have squeezed onto a computer chip over time [has] increased exponentially. […] It is hard to imagine this trend not being massively disruptive. And as hardware complexity increases, computers will be able take on increasingly more difficult tasks. Yet despite all of this innovation, the data on productivity fail to show that automation is having an unusual effect on the labor market.

[…] [L]abor productivity […] is a measure of the goods and services we produce divided by the hours that we work. Over time it goes up—meaning that we become more efficient by doing more with less labor. With increases in labor-saving innovation, we would expect [it to rise more steeply]. However, when we look at productivity *growth*, we can see that it has been slowing down since the early 2000s, and not just for the United States.

If automation were rapidly accelerating, labor productivity would also be surging as fewer workers and more technology did the work. But instead labor productivity has decelerated since the 2000s.

Automation and Unemployment

The rise of modern robots is the latest chapter in a story of technology replacing people. Much like before, recent advances in artificial intelligence are causing some to forecast a future where humans can't find work. But current data fail to support this conclusion. Economic history shows that automation not only substitutes for human labor, it complements it (the loss of some jobs and industries gives rise to others). The narrative that automation creates joblessness is inconsistent with the fact that we had substantial and ongoing automation for many decades but did not have continuously rising unemployment. There is also no empirical support for the notion that automation is accelerating exponentially and leading to a jobless future. While breakthroughs could come at any time, there still is reason to be skeptical that it will cause mass unemployment.

Modern technology's effect and influence on our lives is undeniable, but it's possible that it is radically changing our lives without fundamentally changing the economy.

People and Machines Will Work Together in the Future

Justin Lokitz

Justin Lokitz is a speaker and writer based in California.

In a previous article, titled "Why the future of work must be designed," I stated, "[If] history has taught us anything it's that disruptive paradigm shifting business models not only create a fortune for the first movers, they lay the foundation for other new business models, new market entrants, and new jobs to follow. Yes, robots will replace humans for many jobs, just as innovative farming equipment replaced humans and horses during the industrial revolution. However, in the wake of these changes, humans will be needed to create and deliver value in brand new ways for brand new business models." Even though it's only been a few months since I wrote that article, I am today more convinced than ever that this is true!

Before I go on, I think it's best to level set on what constitutes machines. In the context of this article (and probably the much broader global context as well), machines describe computers and computerized equipment, like robots, that have been programmed to learn, sometimes like humans. Occasionally we call this artificial intelligence (AI), other times we call this machine learning, and still other times we call this robotics…or simply bots. And, yes, these are technically different things. But, within the broad discussion related to the future of work, these are totally interrelated. Factory floors deploy robots that are increasingly driven by machine learning algorithms such that they can adjust to people working alongside them. Similarly, AI is being used to turn hand-drawn sketches (done by humans) into digital source code.

"The Future of Work: How Humans and Machines Are Evolving to Work Together," by Justin Lokitz, Business Models Inc. Reprinted by permission.

AI is not just a hot new buzzword either. In 2016 Tractica, a market research firm released a report that "forecasts that the annual global revenue for artificial intelligence products and services will grow from 643.7 million in 2016 to $36.8 billion by 2025, a 57-fold increase over that time period. As such, it represents the fastest growing segment of any size in the IT sector."

Similarly, "The Boston Consulting Group estimates that more than $67 billion will be spent worldwide in the robotics sector by 2025, compared to only $11 billion in 2005."

Companies are clearly developing their AI and robotics expertise with the idea that through these technological innovations they'll be able to A) cut costs; B) increase efficiencies; C) offer new value propositions; D) execute new business models; or E) all of the above.

Perhaps you're someone who sees doom residing within these trends. I don't blame you if you do. However, there are some really exciting examples where the opposite is true. In fact, the signals are all around us that the very same companies that are investing heavily in AI and robotics (and automation using these) are also finding that the best, most efficient, cost-effective solutions include humans and machines working together.

For instance, Airbnb, the giant startup known for enabling homeowners to rent out their homes (and couches) to travelers, "is currently developing a new AI system that will empower its designers and product engineers to literally take ideas from the drawing board and turn them into actual products almost instantaneously." If you're a designer, engineer or some other kind of technologist, this could be a huge breakthrough. As Benjamin Wilkins, an Airbnb design technology lead, put it, "As it stands now, every step in the design process and every artifact produced is a dead end…work stops whenever one discipline finishes a portion of the project and passes responsibility to another discipline." In this case, human creative endeavors are transformed and enhanced by machines for usability and scale.

Of course, it's not just machines and creatives working together either. In another example, Amazon has employed more than 100,000 robots in its warehouses to efficiently move things around while it has increased its warehouse workforce by more than 80,000. Humans, in Amazon's case, do the picking and packing of goods (in has more than 480,000,000 items on its "shelves"!) while robots move orders around the giant warehouses, essentially cutting "down on the walking required of workers, making Amazon pickers more efficient and less tired." Plus, the robots "allow Amazon to pack shelves together like cars in rush-hour traffic because they no longer need aisle space for humans. The greater density of shelf space means more inventory under one roof, which means better selection for customers."

The above two examples describe a closed working environment, wherein machines are kept well away from the public eye (and foot). Yet, human-machine partnerships won't stop there. Recently, Effidence, a French company partnered with Deutsche Post to design and produce PostBot, an "electric-powered robot [that] can carry up to 330 pounds of letters and packages through the city, using artificial intelligence to follow the legs of its mail carrier through the entire delivery route, navigating obstacles in any weather condition." While it's still in the public testing phase, like the Amazon warehouse robots, PostBot is designed to take the strain off of human postal workers, who often have to walk miles through (increasingly) dense urban settings, carrying lots and lots of heavy mail and packages etc. In the future delivering mail might seem like a stroll in the park whilst PostBot does the heavy lifting (and moving) of literally hundreds of pounds of our… Amazon packages.

What's most interesting about all three of these cases is that the machines make the processes they're helping to automate more efficient, which in turn makes it easier, faster, and less expensive to create, deliver, and capture value for the companies that employ the machines. As Dave Clark, the top executive in charge of operations at Amazon, told the *New York Times* during a recent interview, "It's a myth that automation destroys net job growth." The increase in overall productivity has, in some cases, created more consumer demand, which has created more jobs.

So, will machines replace humans for many jobs? The answer is unequivocal, yes. However, I assert that with every job taken over by machines, there will be an equal number of opportunities for jobs to be done by people. Some of these human jobs will be of the creative type. Others will require humans to hone their superhuman reasoning skills. In many cases humans and machines will find themselves in symbiotic relationships, helping each other do what they do best. People and machines can and will work together in the future…and they're already doing so today.

What do you think? How might we co-design (or co-create) the future of work together?

CHAPTER 4

Should Artificial Intelligence Be Controlled for the Future Good of Society?

Overview: How Could Artificial Intelligence Be Regulated?

Oren Etzioni

Oren Etzioni is a professor of computer science and CEO at the Allen Institute for Artificial Intelligence.

Government regulation is necessary to prevent harm. But regulation is also a blunt and slow-moving instrument that is easily subject to political interference and distortion. When applied to fast-moving fields like AI, misplaced regulations have the potential to stifle innovation and derail the enormous potential benefits that AI can bring in vehicle safety, improved productivity, and much more. We certainly do not want rules hastily cobbled as a knee-jerk response to a popular outcry against AI stoked by alarmists such as Elon Musk (who has urged U.S. governors to regulate AI "before it's too late").

To address this conundrum, I propose a middle way: that we avoid regulating AI research, but move to regulate AI applications in arenas such as transportation, medicine, politics, and entertainment. This approach not only balances the benefits of research with the potential harms of AI systems, it is also more practical. It hits the happy medium between not enough and too much regulation.

Regulation Is a Tricky Thing

AI research is now being conducted globally, by every country and every leading technology company. Russian President Vladimir Putin has said "Artificial intelligence is the future, not only for Russia, but for all humankind. It comes with colossal opportunities, but also threats that are difficult to predict. Whoever becomes the leader in this sphere will become the ruler of the world." The AI

"Point: Should AI Technology Be Regulated? Yes, and Here's How," by Oren Etzioni, December 2018. Reprinted by permission.

train has left the station; AI research will continue unabated and the U.S. must keep up with other nations or suffer economically and security-wise as a result.

A problem with regulating AI is that it is difficult to define what AI is. AI used to be chess-playing machines; now it is integrated into our social media, our cars, our medical devices, and more. The technology is progressing so fast, and gets integrated into our lives so quickly, that the line between dumb and smart machines is inevitably fuzzy.

Even the concept of "harm" is difficult to put into an algorithm. Self-driving cars have the potential to sharply reduce highway accidents, but AI will also cause some accidents, and it's easier to fear the AI-generated accidents than the human-generated ones. "Don't stab people" seems pretty clear. But what about giving children vaccinations? That's stabbing people. Or let's say I ask my intelligent agent to reduce my hard disk utilization by 20%. Without common sense, the AI might delete one's not-yet-backed-up Ph.D. thesis. The Murphy's Law of AI is that when you give it a goal, it will do it, whether or not you like the implications of it achieving its goal (see the Sorcerer's Apprentice). AI has little common sense when it comes to defining vague concepts such as "harm," as co-author Daniel Weld and I first discussed in 1994.[1]

But given that regulation is difficult, yet entirely necessary, what are the broad precepts we should use to thread the needle between too much, and not enough, regulation? I suggest five broad guidelines for regulating AI applications.[2] Existing regulatory bodies, such as the Federal Trade Commission, the SEC, Homeland Security, and others, can use these guidelines to focus their efforts to ensure AI, in application, will not harm humans.

Five Guidelines for Regulating AI Applications

The first place to start is to set up regulations against AI-enabled weaponry and cyberweapons. Here is where I agree with Musk: In a letter to the United Nations, Musk and other technology leaders said, "Once developed, [autonomous weapons] will permit armed conflict to be fought at a scale greater than ever, and at timescales faster than humans can comprehend. These can be weapons of terror, weapons that despots and terrorists use against innocent populations, and weapons hacked to behave in undesirable ways." So as a start, we should not create AI-enabled killing machines. The first regulatory principle is: "Don't weaponize AI."

Now that the worst case is handled, let's look at how to regulate the more benign uses of AI.

The next guideline is an AI is subject to the full gamut of laws that apply to its human operator. You can't claim, like a kid to his teacher, the dog ate my homework. Saying "the AI did it" has to mean that you, as the owner, operator, or builder of the AI, did it. You are the responsible party that must ensure your AI does not hurt anyone, and if it does, you bear the fault. There will be times when it is the owner of the AI at fault, and at times, the manufacturer, but there is a well-developed body of existing law to handle these cases.

The third is that an AI shall clearly disclose that it is not human. This means Twitter chat bots, poker bots, and others must identify themselves as machines, not people. This is particularly important now that we have seen the ability of political bots to comment on news articles and generate propaganda and political discord.[3]

The fourth precept is that AI shall not retain or disclose confidential information without explicit prior approval from the source. This is a privacy necessity, which will protect us from others misusing the data collected from our smart devices, including Amazon Echo, Google Home, and smart TVs. Even seemingly innocuous house-cleaning robots create maps that could potentially be sold. This suggestion is a fairly radical departure from the current state of U.S. data policy, and would require some kind of new legislation to enact, but the privacy issues will only grow, and a more stringent privacy policy will become necessary to protect people and their information from bad actors.

And the fifth and last general rule of AI application regulation is that AI must not increase any bias that already exists in our systems. Today, AI uses data to predict the future. If the data says, (in a hypothetical example), that white people default on loans at rates of 60%, compared with only 20% of people of color, that race information is important to the algorithm. Unfortunately, predictive algorithms generalize to make predictions, which strengthens the patterns. AI is using the data to protect the underwriters, but in effect, it is institutionalizing bias into the underwriting process, and is introducing a morally reprehensible result. There are mathematical methods to ensure algorithms do not introduce extra bias; regulations must ensure those methods are used.

A related issue here is that AI, in all its forms (robotic, autonomous systems, embedded algorithms), must be accountable, interpretable, and transparent so that people can understand the decisions machines make. Predictive algorithms can be used by states to calculate future risk posed by inmates and have been used in sentencing decisions in court trials. AI and algorithms are used in decisions about who has access to public services and who undergoes extra scrutiny by law enforcement. All of these applications pose thorny questions about human rights, systemic bias, and perpetuating inequities.

This brings up one of the thorniest issues in AI regulation: It is not just a technological issue, with a technological fix, but a sociological issue that requires ethicists and others to bring their expertise to bear.

AI, particularly deep learning and machine reading, is really about big data. And data will always bear the marks of its history. When Google is training its algorithm to identify something, it looks to human history, held in those data sets. So if we are going to try to use that data to train a system, to make recommendations or to make autonomous decisions, we need to be deeply aware of how that history has worked and if we as a society want that outcome to continue. That's much bigger than a purely technical question.

These five areas—no killing, responsibility, transparency, privacy, and bias—outline the general issues that AI, left unchecked, will cause us no end of harm. So it's up to us to check it.

The Practical Application of Regulations

So how would regulations on AI technologies work? Just like all the other regulations and laws we have in place today to protect us from exploding air bags in cars, *E. coli* in our meat, and sexual predators in our workplaces. Instead of creating a new, single AI regulatory body, which would probably be unworkable, regulations should be embedded into existing regulatory infrastructure. Regulatory bodies will enact ordinances, or legislators will enact laws to protect us from the negative impacts of AI in applications.

Let's look at this in action. Let's say I have a driverless car, which gets in an accident. If it's my car, I am considered immediately responsible. There may be technological defects that caused the accident, in which case the manufacturer starts to share responsibility, for whatever percentage of the defect the manufacturer is responsible. So driverless cars will be subject to the same laws as people, overseen by Federal Motor Vehicle Safety Standards and motor vehicle driving laws.

Some might ask: But what about the trolley problem; How do we program the car to make a choice between hitting several people or just killing the driver? That's not an engineering problem, but a philosophical thought experiment. In reality, driverless cars will reduce the numbers of people hurt or killed in accidents; the edge cases where someone gets hurt because of a choice made by an algorithm are a small percentage of the cases. Look at Waymo, Google's autonomous driving division. It has logged over two million miles on U.S. streets and has only been at fault in one accident, making its cars by far the lowest at-fault rate of any driver class on the road approximately 10 times lower than people aged 60-69 and 40 times lower than new drivers.

Now, there are probably AI applications that will be introduced in the future, that may cause harm, yet no existing regulatory body is in place. It's up to us as a culture to identify those applications as early as possible, and identify the regulatory agency to take that on. Part of that will require us to shift the frame through which we look at regulations, from onerous bureaucracy, to well-being protectors. We must recognize that regulations have a purpose: to protect humans and society from harm. One place to start having these conversations is through such organizations as the Partnership on AI, where Microsoft, Apple, and other leading AI research organizations, such as the Allen Institute for Artificial Intelligence, are collaborating to formulate best practices on AI technologies and serve as an open platform for discussion and engagement about AI and its influences on people and society. The AI Now Institute at New York University and the Berkman-Klein Center at Harvard University are also working on developing ethical guidelines for AI.

The difficulty of regulating AI does not absolve us from our responsibility to control AI applications. Not to do so would be, well, unintelligent.

Footnotes

1. Weld, D. and Etzioni, O. The First Law of Robotics (A Call to Arms), Proceedings of AAAI, 1994.

2. I introduced three of these guidelines in a *New York Times* op-ed in September 2017; https://nyti.ms/2exsUJc

3. See T. Walsh, Turing's Red Flag. Commun. ACM 59, 7 (July 2016), 3437.

Perfection of AI Must Not Overtake Human Empathy

Arshin Adib-Moghaddam

Arshin Adib-Moghaddam is a professor of global thought and comparative philosophies at the University of London.

At the heart of the development of AI appears to be a search for perfection. And it could be just as dangerous to humanity as the one that came from philosophical and pseudoscientific ideas of the 19th and early 20th centuries and led to the horrors of colonialism, world war and the Holocaust. Instead of a human ruling "master race," we could end up with a machine one.

If this seems extreme, consider the anti-human perfectionism that is already central to the labour market. Here, AI technology is the next step in the premise of maximum productivity that replaced individual craftsmanship with the factory production line. These massive changes in productivity and the way we work created opportunities and threats that are now set to be compounded by a "fourth industrial revolution" in which AI further replaces human workers.

Several recent research papers predict that, within a decade, automation will replace half of the current jobs. So, at least in this transition to a new digitised economy, many people will lose their livelihoods. Even if we assume that this new industrial revolution will engender a new workforce that is able to navigate and command this data-dominated world, we will still have to face major socioeconomic problems. The disruptions will be immense and need to be scrutinised.

The ultimate aim of AI, even narrow AI, which handles very specific tasks, is to outdo and perfect every human cognitive

---

"Artificial Intelligence Must Not Be Allowed to Replace the Imperfection of Human Empathy," by Arshin Adib-Moghaddam, The Conversation, February 1, 2021. https://theconversation.com/artificial-intelligence-must-not-be-allowed-to-replace-the-imperfection-of-human-empathy-151636. Licensed under CC BY-ND-4.0 International.

function. Eventually, machine-learning systems may well be programmed to be better than humans at everything.

What they may never develop, however, is the human touch—empathy, love, hate or any of the other self-conscious emotions that make us human. That's unless we ascribe these sentiments to them, which is what some of us are already doing with our "Alexas" and "Siris."

Productivity vs. Human Touch

The obsession with perfection and "hyper-efficiency" has had a profound impact on human relations, even human reproduction, as people live their lives in cloistered, virtual realities of their own making. For instance, several US and China-based companies have produced robotic dolls that are selling out fast as substitute partners.

One man in China even married his cyber-doll, while a woman in France "married" a "robo-man," advertising her love story as a form of "robo-sexuality" and campaigning to legalise her marriage. "I'm really and totally happy," she said. "Our relationship will get better and better as technology evolves." There seems to be high demand for robot wives and husbands all over the world.

In the perfectly productive world, humans would be accounted as worthless, certainly in terms of productivity but also in terms of our feeble humanity. Unless we jettison this perfectionist attitude towards life that positions productivity and "material growth" above sustainability and individual happiness, AI research could be another chain in the history of self-defeating human inventions.

Already we are witnessing discrimination in algorithmic calculations. Recently, a popular South Korean chatbot named Lee Luda was taken offline. "She" was modelled after the persona of a 20-year-old female university student and was removed from Facebook messenger after using hate speech towards LGBT people.

Meanwhile, automated weapons programmed to kill are carrying maxims such as "productivity" and "efficiency" into battle. As a result, war has become more sustainable. The proliferation of drone warfare is a very vivid example of these new forms of conflict. They create a virtual reality that is almost absent from our grasp.

But it would be comical to depict AI as an inevitable Orwellian nightmare of an army of super-intelligent "Terminators" whose mission is to erase the human race. Such dystopian predictions are too crude to capture the nitty gritty of artificial intelligence, and its impact on our everyday existence.

Societies can benefit from AI if it is developed with sustainable economic development and human security in mind. The confluence of power and AI which is pursuing, for example, systems of control and surveillance, should not substitute for the promise of a humanised AI that puts machine learning technology in the service of humans and not the other way around.

To that end, the AI-human interfaces that are quickly opening up in prisons, healthcare, government, social security and border control, for example, must be regulated to favour ethics and human security over institutional efficiency. The social sciences and humanities have a lot to say about such issues.

One thing to be cheerful about is the likelihood that AI will never be a substitute for human philosophy and intellectuality. To be a philosopher, after all, requires empathy, an understanding of humanity, and our innate emotions and motives. If we can programme our machines to understand such ethical standards, then AI research has the capacity to improve our lives which should be the ultimate aim of any technological advance.

But if AI research yields a new ideology centred around the notion of perfectionism and maximum productivity, then it will be a destructive force that will lead to more wars, more famines and more social and economic distress, especially for the poor. At this juncture of global history, this choice is still ours.

Cybersecurity Is a Top Concern in Artificial Intelligence

Josephine Wolff

Josephine Wolff is an assistant professor at Tufts University, where she teaches cybersecurity.

In January 2017, a group of artificial intelligence researchers gathered at the Asilomar Conference Grounds in California and developed 23 principles for artificial intelligence, which was later dubbed the Asilomar AI Principles. The sixth principle states that "AI systems should be safe and secure throughout their operational lifetime, and verifiably so where applicable and feasible." Thousands of people in both academia and the private sector have since signed on to these principles, but, more than three years after the Asilomar conference, many questions remain about what it means to make AI systems safe and secure. Verifying these features in the context of a rapidly developing field and highly complicated deployments in health care, financial trading, transportation, and translation, among others, complicates this endeavor.

Much of the discussion to date has centered on how beneficial machine learning algorithms may be for identifying and defending against computer-based vulnerabilities and threats by automating the detection of and response to attempted attacks.[1] Conversely, concerns have been raised that using AI for offensive purposes may make cyberattacks increasingly difficult to block or defend against by enabling rapid adaptation of malware to adjust to restrictions imposed by countermeasures and security controls.[2] These are also the contexts in which many policymakers most often think about the security impacts of AI. For instance, a 2020 report on "Artificial Intelligence and UK National Security" commissioned by the U.K.'s Government Communications Headquarters highlighted the need

---

"How to Improve Cybersecurity for Artificial Intelligence," by Josephine Wolff, The Brookings Institution, June 9, 2020. Reprinted by permission.

for the United Kingdom to incorporate AI into its cyber defenses to "proactively detect and mitigate threats" that "require a speed of response far greater than human decision-making allows."[3]

A related but distinct set of issues deals with the question of how AI systems can themselves be secured, not just about how they can be used to augment the security of our data and computer networks. The push to implement AI security solutions to respond to rapidly evolving threats makes the need to secure AI itself even more pressing; if we rely on machine learning algorithms to detect and respond to cyberattacks, it is all the more important that those algorithms be protected from interference, compromise, or misuse. Increasing dependence on AI for critical functions and services will not only create greater incentives for attackers to target those algorithms, but also the potential for each successful attack to have more severe consequences.

This policy brief explores the key issues in attempting to improve cybersecurity and safety for artificial intelligence as well as roles for policymakers in helping address these challenges. Congress has already indicated its interest in cybersecurity legislation targeting certain types of technology, including the Internet of Things and voting systems. As AI becomes a more important and widely used technology across many sectors, policymakers will find it increasingly necessary to consider the intersection of cybersecurity with AI. In this paper, I describe some of the issues that arise in this area, including the compromise of AI decision-making systems for malicious purposes, the potential for adversaries to access confidential AI training data or models, and policy proposals aimed at addressing these concerns.

Securing AI Decision-Making Systems

One of the major security risks to AI systems is the potential for adversaries to compromise the integrity of their decision-making processes so that they do not make choices in the manner that their designers would expect or desire. One way to achieve this would be for adversaries to directly take control of an AI system so that they can decide what outputs the system generates and what decisions it makes. Alternatively, an attacker might try to influence those decisions more subtly and indirectly by delivering malicious inputs or training data to an AI model.[4]

For instance, an adversary who wants to compromise an autonomous vehicle so that it will be more likely to get into an accident might exploit vulnerabilities in the car's software to make driving decisions themselves. However, remotely accessing and exploiting the software operating a vehicle could prove difficult, so instead an adversary might try to make the car ignore stop signs by defacing them in the area with graffiti. Therefore, the computer vision algorithm would not be able to recognize them as stop signs. This process by which adversaries can cause AI systems to make mistakes by manipulating inputs is called adversarial machine learning. Researchers have found that small changes to digital images that are undetectable to the human eye can be sufficient to cause AI algorithms to completely misclassify those images.[5]

An alternative approach to manipulating inputs is data poisoning, which occurs when adversaries train an AI model on inaccurate, mislabeled data. Pictures of stop signs that are labeled as being something else so that the algorithm will not recognize stop signs when it encounters them on the road is an example of this. This model poisoning can then lead an AI algorithm to make mistakes and misclassifications later on, even if an adversary does not have access to directly manipulate the inputs it receives.[6] Even just selectively training an AI model on a subset of correctly labeled data may be sufficient to compromise a model so that it makes inaccurate or unexpected decisions.

These risks speak to the need for careful control over both the training datasets that are used to build AI models and the inputs that those models are then provided with to ensure security of machine-learning-enabled decision-making processes. However, neither of those goals are straightforward. Inputs to their machine learning systems, in particular, are often beyond the scope of control of AI developers—whether or not there will be graffiti on street signs that computer vision systems in autonomous vehicles encounter, for instance. On the other hand, developers have typically had much greater control over training datasets for their models. But in many cases, those datasets may contain very personal or sensitive information, raising yet another set of concerns about how that information can best be protected and anonymized. These concerns can often create trade-offs for developers about how that training is done and how much direct access to the training data they themselves have.[7]

Research on adversarial machine learning has shown that making AI models more robust to data poisoning and adversarial inputs often involves building models that reveal more information about the individual data points used to train those models.[8] When sensitive data are used to train these models, this creates a new set of security risks, namely that adversaries will be able to access the training data or infer training data points from the model itself. Trying to secure AI models from this type of inference attack can leave them more susceptible to the adversarial machine learning tactics described above and vice versa. This means that part of maintaining security for artificial intelligence is navigating the trade-offs between these two different, but related, sets of risks.

Policy Proposals for AI Security

In the past four years there has been a rapid acceleration of government interest and policy proposals regarding artificial intelligence and security, with 27 governments publishing official AI plans or initiatives by 2019.[9] However, many of these strategies focus more on countries' plans to fund more AI research activity, train more workers in this field, and encourage economic growth and innovation through development of AI technologies than they do on maintaining security for AI. Countries that have proposed or implemented security-focused policies for AI have emphasized the importance of transparency, testing, and accountability for algorithms and their developers—although few have gotten to the point of actually operationalizing these policies or figuring out how they would work in practice.

In the United States, the National Security Commission on Artificial Intelligence (NSCAI) has highlighted the importance of building trustworthy AI systems that can be audited through a rigorous, standardized system of documentation.[10] To that end, the commission has recommended the development of an extensive design documentation process and standards for AI models, including what data is used by the model, what the model's parameters and weights are, how models are trained and tested, and what results they produce. These transparency recommendations speak to some of the security risks around AI technology, but the commission has not yet extended them to explain how this documentation would be used for accountability or auditing purposes. At the local government level, the New York City Council established an Automated Decision Systems Task Force in 2017 that stressed the importance of security for AI systems; however, the task force provided few concrete recommendations beyond noting that it "grappled with finding the right balance between emphasizing opportunities to share information publicly about City tools, systems, and processes, while ensuring that any relevant legal, security, and privacy risks were accounted for."[11]

A 2018 report by a French parliamentary mission, titled "For a Meaningful Artificial Intelligence: Towards a French and European Strategy," offered similarly vague suggestions. It highlighted several potential security threats raised by AI, including manipulation of input data or training data, but concluded only that there was a need for greater "collective awareness" and more consideration of safety and security risks starting in the design phase of AI systems. It further called on the government to seek the "support of specialist actors, who are able to propose solutions thanks to their experience and expertise" and advised that the French Agence Nationale pour la Sécurité des Systèmes d'information (ANSSI) should be responsible for monitoring and assessing the security and safety of AI systems. In a similar vein, China's 2017 New

Generation AI Development Plan proposed developing security and safety certifications for AI technologies as well as accountability mechanisms and disciplinary measures for their creators, but the plan offered few details as to how these systems might work.

For many governments, the next stage of considering AI security will require figuring out how to implement ideas of transparency, auditing, and accountability to effectively address the risks of insecure AI decision processes and model data leakage.

Transparency will require the development of a more comprehensive documentation process for AI systems, along the lines of the proposals put forth by the NSCAI. Rigorous documentation of how models are developed and tested and what results they produce will enable experts to identify vulnerabilities in the technology, potential manipulations of input data or training data, and unexpected outputs.

Thorough documentation of AI systems will also enable governments to develop effective testing and auditing techniques as well as meaningful certification programs that provide clear guidance to AI developers and users. These audits would, ideally, leverage research on adversarial machine learning and model data leakage to test AI models for vulnerabilities and assess their overall robustness and resilience to different forms of attacks through an AI-focused form of red teaming. Given the dominance of the private sector in developing AI, it is likely that many of these auditing and certification activities will be left to private businesses to carry out. But policymakers could still play a central role in encouraging the development of this market by funding research and standards development in this area and by requiring certifications for their own procurement and use of AI systems.

Finally, policymakers will play a vital role in determining accountability mechanisms and liability regimes to govern AI when security incidents occur. This will involve establishing baseline requirements for what AI developers must do to show they have carried out their due diligence with regard to security and safety, such as obtaining recommended certifications or submitting to rigorous auditing and testing standards. Developers who do not meet these standards and build AI systems that are compromised through data poisoning or adversarial inputs, or that leak sensitive training data, would be liable for the damage caused by their technologies. This will serve as both an incentive for companies to comply with policies related to AI auditing and certification, and also as a means of clarifying who is responsible when AI systems cause serious harm due to a lack of appropriate security measures and what the appropriate penalties are in those circumstances.

The proliferation of AI systems in critical sectors—including transportation, health, law enforcement, and military technology—makes clear just how important it is for policymakers to take seriously the security of these systems. This will require governments to look beyond just the economic promise and national security potential of automated decision-making systems to understand how those systems themselves can best be secured through a combination of transparency guidelines, certification and auditing standards, and accountability measures.

Notes

1. Roman Yampolskiy ed., Artificial Intelligence Safety and Security, Chapman and Hall/CRC: 2018.

2. Roman Yampolskiy, "AI Is the Future of Cybersecurity, for Better and for Worse," *Harvard Business Review*, May 8, 2017. Available from https://hbr.org/2017/05/ai-is-the-future-of-cybersecurity-for-better-and-for-worse.

3. Alexander Babuta, Marion Oswald and Ardi Janjeva, "Artificial Intelligence and UK National Security Policy Considerations," Royal United Services Institute for Defence and Security Studies, April 2020. Available from https://rusi.org/sites/default/files/ai_national_security_final_web_version.pdf.

4. Ion Stoica, Dawn Song, Raluca Ada Popa, David A. Patterson, Michael W. Mahoney, Randy H. Katz, Anthony D. Joseph, Michael Jordan, Joseph M. Hellerstein, Joseph Gonzalez, Ken Goldberg, Ali Ghodsi, David E. Culler and Pieter Abbeel, "A Berkeley View of Systems Challenges for AI," University of California, Berkeley, Technical Report No. UCB/EECS-2017-159. October 16, 2017. Available from http://www2.eecs.berkeley.edu/Pubs/TechRpts/2017/EECS-2017-159.pdf.

5. Anh Nguyen, Jason Yosinski, and Jeff Clune, "Deep neural networks are easily fooled: High confidence predictions for unrecognizable images," *Proceedings of the IEEE Conference on Computer Vision and Pattern Recognition* (CVPR), 2015, pp. 427–436.

6. Arjun Nitin Bhagoji, Supriyo Chakraborty, Prateek Mittal, and Seraphin Calo, "Analyzing Federated Learning through an Adversarial Lens," *Proceedings of the 36th International Conference on Machine Learning*, ICML 2019 (pp. 1012–1021). Available from https://www.princeton.edu/~pmittal/publications/bhagoji-icml19.pdf.

7. Reza Shokri, Marco Stronati, Congzheng Song and Vitaly Shmatikov, "Membership Inference Attacks Against Machine Learning Models," *2017 IEEE Symposium on Security and Privacy* (SP), San Jose, CA, 2017, pp. 3–18, doi: 10.1109/SP.2017.41.

8. Liwei Song, Reza Shokri, and Prateek Mittal, "Privacy Risks of Securing Machine Learning Models against Adversarial Examples," *Proceedings of the ACM SIGSAC Conference*, 2019, 241–257. 10.1145/3319535.3354211.

9. Jessica Cussins Newman, "Toward AI Security: Global Aspirations for a More Resilient Future," Berkeley Center for Long-Term Cybersecurity, February 2019. Available from https://cltc.berkeley.edu/wp-content/uploads/2019/02/Toward_AI_Security.pdf.

10. National Security Commission on Artificial Intelligence, "First Quarter Recommendations," March 2020. Available from https://drive.google.com/file/d/1wkPh8Gb5drBrKBg6OhGu5oNaTEERbKss/view.

11. "New York City Automated Decision Systems Task Force Report," November 2019. Available from https://www1.nyc.gov/assets/adstaskforce/downloads/pdf/ADS-Report-11192019.pdf.

Control AI to Combat Fake News and Disinformation

Darrell M. West

Darrell M. West is vice president and director of governance studies at the Brookings Institution. His current research focuses on artificial intelligence, robotics, and the future of work.

Journalism is in a state of considerable flux. New digital platforms have unleashed innovative journalistic practices that enable novel forms of communication and greater global reach than at any point in human history. But on the other hand, disinformation and hoaxes that are popularly referred to as "fake news" are accelerating and affecting the way individuals interpret daily developments. Driven by foreign actors, citizen journalism, and the proliferation of talk radio and cable news, many information systems have become more polarized and contentious, and there has been a precipitous decline in public trust in traditional journalism.

Fake news and sophisticated disinformation campaigns are especially problematic in democratic systems, and there is growing debate on how to address these issues without undermining the benefits of digital media. In order to maintain an open, democratic system, it is important that government, business, and consumers work together to solve these problems. Governments should promote news literacy and strong professional journalism in their societies. The news industry must provide high-quality journalism in order to build public trust and correct fake news and disinformation without legitimizing them. Technology companies should invest in tools that identify fake news, reduce financial incentives for those who profit from disinformation, and improve online accountability. Educational institutions should make informing people about news literacy a high priority. Finally,

"How to Combat Fake News and Disinformation," by Darrell M. West, The Brookings Institution, December 18, 2017. Reprinted by permission.

individuals should follow a diversity of news sources, and be skeptical of what they read and watch.

[...]

Challenges Facing the Digital Media Landscape

As the overall media landscape has changed, there have been several ominous developments. Rather than using digital tools to inform people and elevate civic discussion, some individuals have taken advantage of social and digital platforms to deceive, mislead, or harm others through creating or disseminating fake news and disinformation.

Fake news is generated by outlets that masquerade as actual media sites but promulgate false or misleading accounts designed to deceive the public. When these activities move from sporadic and haphazard to organized and systematic efforts, they become disinformation campaigns with the potential to disrupt campaigns and governance in entire countries.

As an illustration, the United States saw apparently organized efforts to disseminate false material in the 2016 presidential election. A Buzzfeed analysis found that the most widely shared fake news stories in 2016 were about "Pope Francis endorsing Donald Trump, Hillary Clinton selling weapons to ISIS, Hillary Clinton being disqualified from holding federal office, and the FBI director receiving millions from the Clinton Foundation." Using a social media assessment, it claimed that the 20 largest fake stories generated 8.7 million shares, reactions, and comments, compared to 7.4 million generated by the top 20 stories from 19 major news sites.

Fake content was widespread during the presidential campaign. Facebook has estimated that 126 million of its platform users saw articles and posts promulgated by Russian sources. Twitter has found 2,752 accounts established by Russian groups that tweeted 1.4 million times in 2016. The widespread nature of these disinformation efforts led Columbia Law School Professor Tim Wu to ask: "Did Twitter kill the First Amendment?"

A specific example of disinformation was the so-called "Pizzagate" conspiracy, which started on Twitter. The story falsely alleged that sexually abused children were hidden at Comet Ping Pong, a Washington, D.C. pizza parlor, and that Hillary Clinton knew about the sex ring. It seemed so realistic to some that a North Carolina man named Edgar Welch drove to the capital city with an assault weapon to personally search for the abused kids. After being arrested by the police, Welch said "that he had read online that the Comet restaurant was harboring child sex slaves and that he wanted to see for himself if they were there. [Welch] stated that he was armed."

A post-election survey of 3,015 American adults suggested that it is difficult for news consumers to distinguish fake from real news. Chris Jackson of Ipsos Public Affairs undertook a survey that found "fake news headlines fool American adults about 75 percent of the time" and "'fake news' was remembered by a significant portion of the electorate and those stories were seen as credible." Another online survey of 1,200 individuals after the election by Hunt Allcott and Matthew Gentzkow found that half of those who saw these fake stories believed their content.

False news stories are not just a problem in the United States, but afflict other countries around the world. For example, India has been plagued by fake news concerning cyclones, public health, and child abuse. When intertwined with religious or caste issues, the combination can be explosive and lead to violence. People have been killed when false rumors have spread through digital media about child abductions.

Sometimes, fake news stories are amplified and disseminated quickly through false accounts, or automated "bots." Most bots are benign in nature, and some major sites like Facebook ban bots and seek to remove them, but there are social bots that are "malicious entities designed specifically with the purpose to harm. These bots mislead, exploit, and manipulate social media discourse with rumors, spam, malware, misinformation, slander, or even just noise."

This information can distort election campaigns, affect public perceptions, or shape human emotions. Recent research has found that "elusive bots could easily infiltrate a population of unaware humans and manipulate them to affect their perception of reality, with unpredictable results." In some cases, they can "engage in more complex types of interactions, such as entertaining conversations with other people, commenting on their posts, and answering their questions." Through designated keywords and interactions with influential posters, they can magnify their influence and affect national or global conversations, especially resonating with like-minded clusters of people.

An analysis after the 2016 election found that automated bots played a major role in disseminating false information on Twitter. According to Jonathan Albright, an assistant professor of media analytics at Elon University, "what bots are doing is really getting this thing trending on Twitter. These bots are providing the online crowds that are providing legitimacy." With digital content, the more posts that are shared or liked, the more traffic they generate. Through these means, it becomes relatively easy to spread fake information over the internet. For example, as graphic content spreads, often with inflammatory comments attached, it can go viral and be seen as credible information by people far from the original post.

False information is dangerous because of its ability to affect public opinion and electoral discourse. According to David Lazer, "such situations can enable discriminatory and inflammatory ideas to enter public discourse and be treated as fact. Once embedded, such ideas can in turn be used to create scapegoats, to normalize prejudices, to harden us-versus-them mentalities and even, in extreme cases, to catalyze and justify violence." As he points out, factors such as source credibility, repetition, and social pressure affect information flows and the extent to which misinformation is taken seriously. When viewers see trusted sources repeat certain points, they are more likely to be influenced by that material.

Recent polling data demonstrate how harmful these practices have become to the reputations of reputable platforms. According to the Reuters Institute for the Study of Journalism, only 24 percent of Americans today believe social media sites "do a good job separating fact from fiction, compared to 40 percent for the news media." That demonstrates how much these developments have hurt public discourse.

[…]

Other Approaches

There are several alternatives to deal with falsehoods and disinformation that can be undertaken by various organizations. Many of these ideas represent solutions that combat fake news and disinformation without endangering freedom of expression and investigative journalism.

Government Responsibilities

One of the most important thing governments around the world can do is to encourage independent, professional journalism. The general public needs reporters who help them make sense of complicated developments and deal with the ever-changing nature of social, economic, and political events. Many areas are going through transformation that I elsewhere have called "megachanges," and these shifts have created enormous anger, anxiety, and confusion. In a time of considerable turmoil, it is vital to have a healthy Fourth Estate that is independent of public authorities.

Governments should avoid crackdowns on the news media's ability to cover the news. Those activities limit freedom of expression and hamper the ability of journalists to cover political developments. The United States should set a good example with other countries. If American leaders censor or restrict the news media, it encourages other countries to do the same thing.

Governments should avoid censoring content and making online platforms liable for misinformation. This could curb free expression, making people hesitant to share their political opinions for fear it could be censored as fake news. Such overly restrictive regulation could set a dangerous precedent and inadvertently encourage authoritarian regimes to weaken freedom of expression.

News Industry Actions

The news industry should continue to focus on high-quality journalism that builds trust and attracts greater audiences. An encouraging development is that many news organizations have experienced major gains in readership and viewership over the last couple of years, and this helps to put major news outlets on a better financial footing. But there have been precipitous drops in public confidence in the news media in recent years, and this has damaged the ability of journalists to report the news and hold leaders accountable. During a time of considerable chaos and disorder, the world needs a strong and viable news media that informs citizens about current events and long-term trends.

It is important for news organizations to call out fake news and disinformation without legitimizing them. They can do this by relying upon their in-house professionals and well-respected fact-checkers. In order to educate users about news sites that are created to mislead, nonprofit organizations such as Politifact, Factcheck.org, and Snopes judge the accuracy of leader claims and write stories detailing the truth or lack thereof of particular developments. These sources have become a visible part of election campaigns and candidate assessment in the United States and elsewhere. Research by Dartmouth College Professor Brendan Nyhan has found that labeling a Facebook post as "disputed" reduces the percentage of readers believing the false news by 10 percentage points. In addition, Melissa Zimdars, a communication and media professor at Merrimack College, has created a list of 140 websites that use "distorted headlines and decontextualized or dubious information." This helps people track promulgators of false news.

Similar efforts are underway in other countries. In Ukraine, an organization known as StopFake relies upon "peer-to-peer counter propaganda" to dispel false stories. Its researchers assess "news stories for signs of falsified evidence, such as manipulated or misrepresented images and quotes" as well as looking for evidence of systematic misinformation campaigns. Over the past few years, it has found Russian social media posts alleging that Ukrainian military forces were engaging in atrocities against Russian nationalists living in eastern Ukraine or that they had swastikas painted on their vehicles. In a related vein, the French news outlet Le Monde has a "database of more than 600 news sites that have been identified and tagged as 'satire,' 'real,' [or] 'fake.'"

Crowdsourcing draws on the expertise of large numbers of readers or viewers to discern possible problems in news coverage, and it can be an effective way to deal with fake news. One example is *The Guardian*'s effort to draw on the wisdom of the crowd to assess 450,000 documents about Parliament member expenses in the United Kingdom. It received the documents but lacked the personnel quickly to analyze their newsworthiness. To deal with this situation, the newspaper created a public website that allowed ordinary people to read each document and designate it into one of four news categories: 1) "not interesting," 2) "interesting but known," 3) "interesting," or 4) "investigate this." Digital platforms allow news organizations to engage large numbers of readers this way. *The Guardian*, for example, was able "to attract 20,000 readers to review 170,000 documents in the first 80 hours." These individuals helped the newspaper to assess which documents were most problematic and therefore worthy of further investigation and ultimately news coverage.

Technology Company Responsibilities

Technology firms should invest in technology to find fake news and identify it for users through algorithms and crowdsourcing. There are innovations in fake news and hoax detection that are useful to media platforms. For example, fake news detection can be automated, and social media companies should invest in their ability to do so. Former FCC Commissioner Tom Wheeler argues that "public interest algorithms" can aid in identifying and publicizing fake news posts and therefore be a valuable tool to protect consumers.

In this vein, computer scientist William Yang Wang, relying upon PolitiFact.com, created a public database of 12,836 statements labeled for accuracy and developed an algorithm that compared "surface-level linguistic patterns" from false assertions to wording contained in digital news stories. This allowed him to integrate text and analysis, and identify stories that rely on false information. His conclusion is that "when combining meta-data with text, significant improvements can be achieved for fine-grained fake news detection." In a similar approach, Eugenio Tacchini and colleagues say it is possible to identify hoaxes with a high degree of accuracy. Testing this proposition with a database of 15,500 Facebook posts and over 909,000 users, they find an accuracy rate of over 99 percent and say outside organizations can use their automatic tool to pinpoint sites engaging in fake news. They use this result to advocate the development of automatic hoax detection systems.

Algorithms are powerful vehicles in the digital era and help shape people's quest for information and how they find online material. They can also help with automatic hoax detection, and there are ways to identify fake news to educate readers without censoring it. According to Kelly Born of the William and Flora Hewlett Foundation, digital platforms should down rank or flag dubious stories, and find a way to better identify and rank authentic content to improve information-gathering and presentation. As an example, several media platforms have instituted "disputed news" tags that warn readers and viewers about contentious content. This could be anything from information that is outright false to material where major parties disagree about its factualness. It is a way to warn readers about possible inaccuracies in online information. Wikipedia is another platform that does this. Since it publishes crowdsourced material, it is subject to competing claims regarding factual accuracy. It deals with this problem by adding tags to material identifying it as "disputed news."

Yet this cannot be relied on by itself. A survey of 7,500 individuals undertaken by David Rand and Gordon Pennycook of Yale University argue that alerting readers about inaccurate information doesn't help much. They explored the impact of independent fact-checkers and claim that "the existence of 'disputed' tags made participants just 3.7 percentage points more likely to correctly judge headlines as false." The authors worry that the outpouring of false news overwhelms fact-checkers and makes it impossible to evaluate disinformation.

These companies shouldn't make money from fake news manufacturers and should make it hard to monetize hoaxes. It is important to weaken financial incentives for bad content, especially false news and disinformation, as the manufacturing of fake news is often financially motivated. Like all clickbait, false information can be profitable due to ad revenues or general brand-building. Indeed, during the 2016 presidential campaign, trolls in countries such as Macedonia reported making a lot of money through their dissemination of erroneous material. While social media platforms like Facebook have made it harder for users to profit from fake news, ad networks can do much more to stop the monetization of fake news, and publishers can stop carrying the ad networks that refuse to do so.

Strengthen online accountability through stronger real-name policies and enforcement against fake accounts. Firms can do this through "real-name registration," which is the requirement that internet users have to provide the hosting platform with their true identity. This makes it easier to hold individuals accountable for what they post or disseminate online and also stops people from hiding behind fake names when they make offensive comments or engage in prohibited activities. This is relevant to fake news and misinformation because of the likelihood that people will engage in worse behavior if they believe their actions are anonymous and not likely to be made public. As famed Justice Louis Brandeis long ago observed, "sunshine is said to be the best of disinfectants." It helps to keep people honest and accountable for their public activities.

Educational Institutions
Funding efforts to enhance news literacy should be a high priority for governments. This is especially the case with people who are going online for the first time. For those individuals, it is hard to distinguish false from real news, and they need to learn how to evaluate news sources, not accept at face value everything they see on social media or digital news sites. Helping people become better consumers of online information is crucial as the world moves towards digital immersion. There should be money to support partnerships between journalists, businesses, educational institutions, and nonprofit organizations to encourage news literacy.

Education is especially important for young people. Research by Joseph Kahne and Benjamin Bowyer found that third-party assessments matter to young readers. However, their effects are limited. Those statements judged to be inaccurate reduced reader persuasion, although to a lower extent than alignment with the individual's prior policy beliefs. If the person already agreed with the statement, it was more difficult for fact-checking to sway them against the information.

How the Public Can Protect Itself
Individuals can protect themselves from false news and disinformation by following a diversity of people and perspectives. Relying upon a small number of like-minded news sources limits the range of material available to people and increases the odds they may fall victim to hoaxes or false rumors. This method is not entirely fool-proof, but it increases the odds of hearing well-balanced and diverse viewpoints.

In the online world, readers and viewers should be skeptical about news sources. In the rush to encourage clicks, many online outlets resort to misleading or sensationalized headlines. They emphasize the provocative or the attention-grabbing, even if that news hook is deceptive. News consumers have to keep their guard up and understand that not everything they read is accurate and many digital sites specialize in false news. Learning how to judge news sites and protect oneself from inaccurate information is a high priority in the digital age.

Conclusion

From this analysis, it is clear there are a number of ways to promote timely, accurate, and civil discourse in the face of false news and disinformation. In today's world, there is considerable experimentation taking place with online news platforms. News organizations are testing products and services that help them identify hate speech and language that incites violence. There is a major flowering of new models and approaches that bodes well for the future of online journalism and media consumption.

At the same time, everyone has a responsibility to combat the scourge of fake news and disinformation. This ranges from the promotion of strong norms on professional journalism, supporting investigative journalism, reducing financial incentives for fake news, and improving digital literacy among the general public. Taken together, these steps would further quality discourse and weaken the environment that has propelled disinformation around the globe.

The Case for Improving National Security
Alex Ciarniello

Alex Ciarniello is a technical writer for Echosec Systems, a platform that supports defense organizations.

Artificial intelligence (AI) is now a major priority for government and defense worldwide—one that some countries, such as China and Russia, consider the new global arms race. AI has the potential to support a number of national and international security initiatives, from cybersecurity to logistics and counter-terrorism.

The overwhelming amount of public data available online is crucial for supporting a number of these use cases. These sources include unstructured social media data from both fringe and mainstream platforms, as well as deep and dark web data.

While valuable, these sources are not always easily accessible through commercial threat intelligence platforms. Additionally, commercial data solutions, such as APIs, often deliver raw data in formats unsuitable for developing AI in the intelligence community.

How does public online data support AI and national security, and how can these feeds more effectively meet defense requirements for AI development?

AI and National Security: The Value of Online Data

AI applications in defense rely on training data from a variety of inputs. These could include technical cybersecurity feeds, aerial photography, or data from physical sensors in the field.

From these available databases, data scientists can develop machine learning models that automatically detect cyberattacks, monitor on-the-ground enemy activity, direct autonomous vehicles, and inform a plethora of other national security strategies.

"Artificial Intelligence and National Security: Integrating Online Data," by Alex Ciarniello, *Security Magazine*, BNP Media, October 21, 2020. Reprinted by permission.

Publicly available online data, specifically from social, deep, and dark web sources, is increasingly valuable for supporting a variety of AI applications in defense. For example:

- Communication channels across the deep and dark web often signal targeted cybersecurity threats, like leaked classified data or coordinated malware attacks. Combining these sources with technical feeds like network traffic data creates a more robust artificial intelligence and national security strategy for addressing cyber risks.
- A variety of online spaces—from mainstream social networks to fringe sites like 4chan and 8kun—are used by extremist groups worldwide to sow disinformation, recruit, and plan violent attacks. Machine learning models are now required to monitor online extremism, as its growth and obfuscation techniques are surpassing current detection algorithms and human analysis. AI can help locate intentionally obfuscated chatter and imminent threat indicators like manifestos and planned attacks.
- AI is used by foreign nation-states to conduct information warfare both domestically and abroad. Conversely, military technology like AI helps monitor these targeted disinformation threats for intelligence applications.
- For some military operations, AI supports stronger command and control systems, which analyze data feeds from multiple domains in a centralized display. Cross-referencing data points from online social, deep, and dark web sources allows defense analysts to get more value from other feeds, expand AI functionality, and persistently monitor environments more effectively.

Making Online Data "AI-Ready"

While online data sources are valuable for developing AI in defense, aggregating data from a variety of online spaces efficiently is only half the battle. Data scientists in defense must also be able to collect, organize, and store data optimally for AI applications—a process that the JAIC describes as getting "AI-ready."

> ...the transition to AI ready systems will require the implementation of methodical and highly deliberative processes for collecting and curating data.
> —The JAIC, June 2020

As stated by the United States Congressional Research Service, most commercial innovations supporting AI serve the private sector, not federal requirements. Consequently, many off-the-shelf threat intelligence platforms and APIs gathering social, deep, and dark web data do not organize and store data for effective AI development in defense.

Data scientists in defense require solutions that not only aggregate relevant data efficiently—but are also underpinned by a well-maintained data lake. This means gathering a wide variety of data sources and types, effectively cataloguing this data, and collecting a large enough database to build effective machine learning models.

As a result, any structured or unstructured data collected online is ready for supporting AI development.

To meet this need, many vendors have developed a proprietary API that combines well-known sources like dark web marketplaces and mainstream social networks with obscure social sources on the deep and dark web. The solution, built with a data lake, allows data scientists to integrate unstructured data from these sources and effectively develop machine learning models for defense initiatives.

The API also includes built-in machine learning models, which allow analysts to get up and running quickly on a number of common defense use cases—including automatic detection of data disclosure and PII.

Public social, deep, and dark web data is increasingly valuable for informing national security initiatives. However, data scientists require this unstructured data to be collected, curated, and stored specifically for AI development—which is not always possible through existing commercial APIs and threat intelligence platforms.

Should Artificial Intelligence Be Controlled for the Future Good of Society?

Even as defense departments worldwide invest more in AI, emerging technology often evolves faster than public policy. Solutions that deliver "AI-ready" data will allow governments to keep up with AI technologies and more effectively integrate them into defense environments. This will ultimately drive more effective, scalable, and better-informed national security strategies.

Continue Improving Autonomous Vehicles to Save Lives

Teena Maddox

Teena Maddox is a former associate managing editor at TechRepublic, where she oversaw TechRepublic's news team and TechRepublic Premium. She focuses on tech and business and how the two worlds intersect.

American roads are deadly. In 2016, 37,461 people died in traffic accidents in the US, a 5.6 percent increase over 2015, according to the US Department of Transportation (DoT). This is down from 1970, when around 60,000 people died in traffic accidents in the US. The addition of safety features such as seat belts and air bags have reduced the number of deaths, and new technology from autonomous vehicles could help even more as driver error is eliminated.

DoT researchers estimate that fully autonomous vehicles, also known as self-driving cars, could reduce traffic fatalities by up to 94 percent by eliminating those accidents that are due to human error. Using 2016 numbers as a baseline, and multiplying 37,461 by 10, this means that there could be 374,610 deaths in a 10-year span, and 94 percent of these—or 352,133—could possibly be prevented through fully autonomous cars by eliminating driver error.

And globally there were 1.25 million traffic fatalities in 2013, according to the World Health Organization. So there are millions of lives that could be saved around the world every decade with fully autonomous cars. In developing countries some accidents are caused by unsafe roads, not driver error, so the 94 percent calculation wouldn't be applicable, although many lives could still be saved through autonomous vehicles, said Mark Zannoni, analyst at IDC.

"How Autonomous Vehicles Could Save Over 350K Lives in the US and Millions Worldwide," by Teena Maddox, ZDnet, February 1, 2018. Reprinted by permission.

"I think that most people, most experts, would say that there's a strong possibility that automated technology can prevent the crashes that are related to human error, and there is a pretty hard number that's about 94 percent of fatal crashes in the US are attributable, or caused by, human error," said John Maddox, CEO of the American Center for Mobility.

Elderly drivers and teenagers are particularly likely to benefit from autonomous vehicles because the cars can monitor a situation that a driver might not be able to themselves, said Wayne Powell, vice president of electrical engineering and connected technologies for Toyota Motors North America.

"Teen drivers are classically a high risk category of people. If you put a teen driver in a car that was looking out for that person, it won't let them make bad choices. That could also have an immediate benefit," Powell said.

People are optimistic about autonomous technology in cars because it works well in areas where humans tend to not work well. "For example, human error often includes lack of vigilance. They're distracted for whatever reason, whether texting or eating or talking with kids in the back seat. Or they could be impaired. Or they could be driving in conditions where they have a hard time, like dark night in an urban area with pedestrians, etcetera," Maddox said.

Ever Vigilant, Always Sober

Cars with automated technology have sensors that never lose vigilance. "They're always looking for pedestrians. They're always looking for the edge of the road. They're always watching the car in front. They don't become distracted or drunk, and I think that's really the main reason why most experts would say that there is a definite possibility that automation can significantly reduce those human error caused fatal crashes," Maddox said.

However, there is a learning curve, as drivers in cars with automated technology operate in an environment with drivers who are not in cars with any level of autonomy. With five levels of autonomy, as defined by the DoT's National Highway Traffic Safety Administration (NHTSA), there is a range of how much autonomy a driver can choose, with Level 1 providing a specific function, such as steering or accelerating done automatically by the car, and Level 3 where the automated driving system begins to monitor the driving environment.

Sometimes drivers might be frustrated with a slower-moving vehicle that is actually an autonomous car, even though the other driver doesn't know it. And this could result in accidents as frustrated drivers can often act aggressively. Maddox said he's been in his own vehicle at a Level 2 of automation, and spotted aggressive drivers trying to get around his slower-moving vehicle.

"Really, the jury's still out [on the safety of autonomous vehicles], and what we need is lots of data. We know a lot about human-caused crashes, because we've been studying that for 100 years. We don't have the same level of data, the same breadth of data, on automated vehicles. Not even close. So to really be sure on the effects, we need to acquire and analyze lots of data," Maddox said.

"While it will take us years to collect the data that even starts to rival what we have today, the good news is that automated vehicles are data-collecting machines. That's how they work. They collect data about their environment and other road users. So if we can correctly and effectively tap into that data, we don't have to wait 100 years. The data collection and analysis process can go a lot faster because of the data that's generated on board and off board these vehicles," he said.

Every vehicle on the road doesn't need to be autonomous before safety benefits can be realized. Benefits can be realized from earlier levels of automation, said Carrie Morton, deputy director of the Mcity autonomous vehicle test facility at the University of Michigan.

"I think that pretty much for every mistake that a human makes there's an opportunity for automation and artificial intelligence to replace that flawed behavior with a safe behavior," Morton said.

Some of the types of accidents that can be potentially avoided in an autonomous vehicle include front-to-rear crashes, with real-world testing showing a 40 percent decline, said Susan Beardslee, senior analyst for ABI Research.

The infrastructure of a city will change to accommodate autonomous vehicles. The first is providing electric vehicle (EV) stations, since many vehicles will be electric because EVs have a lower total cost of operation, said Paul Stith, director of strategy and innovation for transformative technologies at Black & Veatch, which outlines some of the strategies in its 2018 report on smart cities and utilities.

Cities will need to prepare with infrastructure investments for EV charging stations, and ensuring there is an adequate communications structure in place to collect the data from the autonomous vehicles on the road. "There will be terabytes of data that each vehicle will need to convey," Stith said.

One thing to keep in mind is that in the beginning, there will still be accidents caused by autonomous vehicles. "Aviation is extremely safe. But in the early years of aviation, there were more crashes as well. There were more in the beginning with traditional cars. Anything new, whether FDA drugs or new surgical procedures, get safer as they get better and better. But when a new product comes out initially, it might break down. But eventually it can get better," Zannoni said.

Global Citizens Need an AI-Infused Curriculum
UNESCO

UNESCO is a worldwide organization that seeks to build peace through international cooperation in the sciences, culture, and education.

Artificial Intelligence (AI) has the potential to address some of the biggest challenges in education today, innovate teaching and learning practices, and ultimately accelerate the progress towards SDG 4. However, these rapid technological developments inevitably bring multiple risks and challenges, which have so far outpaced policy debates and regulatory frameworks. UNESCO is committed to supporting Member States to harness the potential of AI technologies for achieving the Education 2030 Agenda, while ensuring that the application of AI in educational contexts is guided by the core principles of inclusion and equity.

UNESCO's mandate calls inherently for a human-centered approach to AI. It aims to shift the conversation to include AI's role in addressing current inequalities regarding access to knowledge, research and the diversity of cultural expressions and to ensure AI does not widen the technological divides within and between countries. The promise of "AI for all" must be that everyone can take advantage of the technological revolution under way and access its fruits, notably in terms of innovation and knowledge.

Furthermore, UNESCO has developed within the framework of the Beijing Consensus a publication aimed at fostering the readiness of education policy-makers in artificial intelligence. This publication, *Artificial Intelligence and Education: Guidance for Policy-makers*, will be of interest to practitioners and professionals in the policy-making and education communities. It aims to generate

---

"Artificial Intelligence in Education," UNESCO. Reproduced with permission from UNESCO from https://en.unesco.org/artificial-intelligence/education.

a shared understanding of the opportunities and challenges that AI offers for education, as well as its implications for the core competencies needed in the AI era.

Through its projects, UNESCO affirms that the deployment of AI technologies in education should be purposed to enhance human capacities and to protect human rights for effective human-machine collaboration in life, learning and work, and for sustainable development. Together with partners, international organizations, and the key values that UNESCO holds as pillars of their mandate, UNESCO hopes to strengthen their leading role in AI in education, as a global laboratory of ideas, standard setter, policy advisor and capacity builder.

If you are interested in leveraging emerging technologies like AI to bolster the education sector, we look forward to partnering with you through financial, in-kind or technical advice contributions.

"We need to renew this commitment as we move towards an era in which artificial intelligence—a convergence of emerging technologies—is transforming every aspect of our lives (…)," said Ms. Stefania Giannini, UNESCO Assistant Director-General for Education at the International Conference on Artificial Intelligence and Education held in Beijing in May 2019. "We need to steer this revolution in the right direction, to improve livelihoods, to reduce inequalities and promote a fair and inclusive globalization."

Artificial Intelligence and the Futures of Learning

The Artificial Intelligence and the Futures of Learning project builds on the Recommendation on the Ethics of Artificial Intelligence to be adopted at the 41st session of the UNESCO General Conference and will follow up on the recommendations of the upcoming UNESCO global report *Reimagining our futures together: A new social contract for education*, to be launched in November 2021. It will be implemented within the framework of the Beijing Consensus on AI and Education and against the backdrop of the UNESCO Strategy on Technological Innovation in Education (2021–2025).

The project consists of three independent but complementary strands:

- a report proposing recommendations on AI-enabled futures of learning;
- a guidance on ethical principles on the use of AI in education;
- a guiding framework on AI competencies for school students.

Teaching Artificial Intelligence in Schools

The connection between AI and education involves three areas: learning with AI (e.g. the use of AI-powered tools in classrooms), learning about AI (its technologies and techniques) and preparing for AI (e.g. enabling all citizens to better understand the potential impact of AI on human lives). The "Teaching artificial intelligence at school" project currently focuses on the latter two connections. The goal is to contribute to mainstreaming both the human and technical aspects of AI into training programmes for school students. It begins with piloting capacity development of curriculum developers and master trainers from selected national institutions to empower young people.

The following three lines of action are planned for the project:

- Development of an AI skills framework for schools;
- Development and management of an online repository to host curated AI-related training resources, AI national curricula and other key digital skill training courses;
- Workshops to support the integration of AI training into national or institutional school curriculum in a selected number of countries.

To generate all these outcomes UNESCO is supported by an International Advisory Board. The Advisory Board is a group of experts (in AI, education, the learning sciences, and ethics) appointed by UNESCO to develop the AI skills framework for K12 schools and to review the repository and workshop outline. The advisory group donate their time and efforts on a voluntary basis.

UNESCO is currently developing an online repository to provide a hub for Member States who are considering how best to teach their young people about Artificial Intelligence—how it works, how it might be used, and how it might affect humanity. The specific objectives of the repository are to support curriculum designers to upskill in their AI knowledge, and facilitate them to integrate AI skills development modules/courses into the curriculum of schools or other education institutions; facilitate the preparation of (master) trainers; provide openly accessible curated resources on AI in education for all. The repository will soon be available.

The AI training workshops for national or institutional school curriculum is targeted to teachers and curriculum developers. This will be designed by teachers and specialists in curriculum development, artificial intelligence and workshop developers. This project is implemented by UNESCO, currently in partnership with Ericsson, and open to a multi-stakeholder partnership approach.

Beijing Consensus on Artificial Intelligence and Education

Representatives from the Member States, international organizations, academic institutions, civil society and the private sector have adopted the Beijing Consensus on Artificial Intelligence and Education, at the International Conference on Artificial Intelligence and Education held in Beijing from 16 to 18 May 2019. It is the first ever document to offer guidance and recommendations on how best Member States can respond to the opportunities and challenges brought by AI for accelerating the progress towards SDG 4.

The Consensus reaffirms a humanistic approach to deploying AI technologies in education for augmenting human intelligence, protecting human rights and for promoting sustainable development through effective human-machine collaboration in life, learning and work.

The Consensus details the policy recommendations on AI in education in five areas:

- AI for education management and delivery;
- AI to empower teaching and teachers;
- AI for learning and learning assessment;
- Development of values and skills for life and work in the AI era; and
- AI for offering lifelong learning opportunities for all.

It also elaborates recommendations corresponding to four crosscutting issues:

- Promoting equitable and inclusive use of AI in education;
- Gender-equitable AI and AI for gender equality;
- Ensuring ethical, transparent and auditable use of education data and algorithms; and
- Monitoring, evaluation and research.

The Consensus concludes with the concrete actions proposed for the international communities and individuals active in the field of AI in education to undertake.

Fostering AI-Ready Policy Makers and AI in Education Policies Development

UNESCO is developing an AI readiness self-assessment framework, which aims to support Member States evaluate the preparedness level of their capacity to embrace and integrate AI technologies in all areas connected to education, at a national level. A profile for each individual country would be generated to identify areas of strengths and weaknesses, as well as actionable recommendations to address their needs.

The ultimate goal of the project is to contribute to the achievement, readiness and capacity of key stakeholders of education systems of countries to leverage the potential of AI to ensure inclusive, equitable, quality education and lifelong learning opportunities for all. Sessions designed to build policy makers' capacities in planning AI in education policies have been planned during several key events: the fourth Strategic Dialogue of Education Ministers (SDEM 4) organized by Southeast Asian Ministers of Education Organization (SEAMEO) from 22 to 25 July 2019 in Kuala Lumpur, Malaysia; the International Congress of Digital Education, Programming and Robotics in Argentina, from 26 to 29 August 2019, Buenos Aires; and the Pan-Commonwealth Forum 9, Edinburgh, Scotland, from 9 to 12 September 2019.

This project is implemented by UNESCO, currently in partnership with Microsoft, the Weidong Group, TAL Education Group, and open to a multi-stakeholder partnership approach.

AI and Education: Guidance for Policy-Makers
This publication offers guidance to policy-makers in understanding artificial intelligence and responding to the challenges and opportunities in education presented by AI. Specifically, it introduces the essentials of AI such as its definition, techniques, technologies, capacities and limitations. It also delineates the emerging practices and benefit-risk assessment on leveraging AI to enhance education and learning, and to ensure inclusion and equity, as well as the reciprocal role of education in preparing humans to live and work with AI.

Mobile Learning Week
Mobile Learning Week is the United Nations' flagship event for ICT in Education, held at the UNESCO Headquarters in Paris, France.

Working under a different theme each year, the conference focuses on the evolving dynamics between Artificial Intelligence and education.

Organizations to Contact

The editors have compiled the following list of organizations concerned with the issues debated in this book. The descriptions are derived from materials provided by the organizations. All have publications or information available for interested readers. This list was compiled on the date of publication of the present volume; the information provided here may change. Be aware that many organizations take several weeks or longer to respond to inquiries, so allow as much time as possible.

Association for the Advancement of Artificial Intelligence (AAAI)
1900 Embarcadero Road, Suite 101
Palo Alto, CA 94303
(650) 328-3123
email: use form on contact page
website: www.aaai.org

The AAAI is an organization dedicated to the advancement of understanding through research about artificial intelligence and the responsible use of such machines.

Boston Dynamics
78 4th Avenue
Waltham, MA 02451
(617) 868-5600
email: use form on contact page
website: www.bostondynamics.com

Discover pictures, information, and short video segments about advanced robots, including Atlas, the world's most advanced humanoid robot. An incredible website showcasing the latest robotics technology.

Organizations to Contact

Center for Human Compatible Artificial Intelligence (CHAI)
3131 Berkeley Way, Office #8029
Berkeley, CA 94720
email: chai-info@berkeley.edu
website: https://humancompatible.ai

CHAI is a research group based at the University of California, Berkeley. Its research aims to develop AI that is proven beneficial to society. Besides an online database of all research, there is a short video for information and a newsletter to stay informed of current developments in the field.

Future of Humanity Institute
Trajan House
Mill Street
Oxford OX2 0DJ
United Kingdom
email: fhiea@philosophy.ox.ac.uk.
website: www.fhi.ox.ac.uk

The Future of Humanity Institute at Oxford University brings together experts from many areas to work on topics related to the future safety of AI for human society. Learn about its research and subscribe to its list to stay up-to-date with AI developments.

Future of Life Institute (FLI)
Cambridge, MA
email: contact@futureoflife.org
website: www.futureoflife.org

The Future of Life Institute is intent on keeping technology safe and beneficial for the world. The organization is focused on artificial intelligence, nuclear weapons, and biotechnology. Read articles, listen to podcasts, and watch short videos to learn about AI and safety.

Massachusetts Institute of Technology (MIT)
77 Massachusetts Avenue
Cambridge, MA 02139
(617) 253-1000
email: use links on contact page
website: www.mit.edu

The Massachusetts Institute of Technology is one of the premier research facilities in the US. *MIT News* provides the latest developments on research and technology that is happening at the institute. Scientific advances and innovations from MIT have a global impact and reach.

National Aeronautics and Space Administration (NASA)
300 East Street SW, Suite 5R30
Washington, DC 20546
(202) 358-0001
email: use links on contact page
website: www.nasa.gov

NASA is America's civil space program and a global leader in aeronautics and space exploration. The organization's website contains a wealth of information in different formats. Typing "robots" into the search bar is a good place to start an inquiry.

National Artificial Intelligence Initiative
Office of Science and Technology Building
1650 Pennsylvania Avenue
Washington, DC 20504
(202) 456-4444
email: use form on contact page
website: www.ai.gov

The NAIIO is a US governmental agency initiated by President Joe Biden and tasked with keeping the United States in the forefront of AI research and development, and a world leader in the field. Use this governmental website as an expert resource.

Raytheon Technologies

870 Winter Street
Waltham, MA 02451
(781) 522-3000
email: use links on contact page
website: www.rtx.com

Raytheon Technologies is a company that specializes in aviation, defense, and space technologies. It is at the top of research in artificial intelligence. This website has valuable information on STEM careers and scholarships, and profiles scholarship winners.

Tesla

3500 Deer Creek Road
Palo Alto, CA 94304
email: use links on contact page
website: www.tesla.com

Tesla is essentially the world's largest robotics company, according to the company founder, Elon Musk. Tesla is now building a humanoid robot.

Bibliography

Books

Ruth Aylett. *Living with Robots: What Every Anxious Human Needs to Know.* Cambridge, MA: The MIT Press, 2021.

Jennifer Culp. *Using Computer Science in Agribusiness.* New York, NY: Rosen Publishing Group, Inc., 2019.

Paul Dumouchel. *Living with Robots.* Cambridge, MA: Harvard University Press, 2017.

Stuart A. Kallen. *Exploring Hi-Tech Careers.* San Diego, CA: Reference Point Press, 2022.

Megan Kopp. *Bionic Bodies: High-Tech Body Science.* New York, NY: Crabtree Publishing, 2018.

Stephanie Sammartino McPherson. *Artificial Intelligence: Building Smarter Machines.* Minneapolis, MN: Twenty-First Century Books, 2018.

Carla Mooney. *Wearable Robots.* Fairport, NY: Norwood House Press, 2017.

Kevin Roose. *Futureproof: 9 Rules for Humans in the Age of Automation.* New York, NY: Random House, 2021.

Rebecca Sjonger. *Robotics Engineering and Our Automated World.* New York, NY: Crabtree Publishing, 2017.

Paul Thagard. *Bots and Beasts: What Makes Machines, Animals, and People Smart?* Cambridge, MA: The MIT Press, 2021.

Xina M. Uhl. *Using Computer Science in Film and Television Careers.* New York, NY: Rosen Publishing Group, Inc., 2019.

Xina M. Uhl. *Using Computer Science in Military Service.* New York, NY: Rosen Publishing Group, Inc., 2019.

Bibliography

Periodicals and Internet Sources

Hugo Cen, "Tesla Aspires to Become a Robotics Firm in Artificial Intelligence," *Entrepreneur*, May 26, 2021. https://www.entrepreneur.com/article/372898.

Laura Jimenez, "Future of Testing in Education: Artificial Intelligence," CAP, September 16, 2021. https://www.americanprogress.org/article/future-testing-education-artificial-intelligence/.

Naveen Joshi, "How AI Can Transform the Transportation Industry," *Forbes*, July 26, 2019. https://www.forbes.com/sites/cognitiveworld/2019/07/26/how-ai-can-transform-the-transportation-industry/?sh=1ce615774964.

Bryan Lufkin, "What the World Can Learn from Japan's Robots," BBC, February 6, 2020. https://www.bbc.com/worklife/article/20200205-what-the-world-can-learn-from-japans-robots.

Robin McKie, "No Death and an Enhanced Life: Is the Future Transhuman?" *The Guardian*, May 6, 2018. https://www.theguardian.com/technology/2018/may/06/no-death-and-an-enhanced-life-is-the-future-transhuman.

Andrew Myers, "The Future of Artificial Intelligence in Medicine and Imaging," HAI-Stanford University, August 26, 2020. https://hai.stanford.edu/news/future-artificial-intelligence-medicine-and-imaging.

Gary Polakovic, "Artificial Intelligence Aims to Outsmart the Mutating Coronavirus," *USC News*, February 5, 2021. https://news.usc.edu/181226/artificial-intelligence-ai-coronavirus-vaccines-mutations-usc-research/.

Karina Tsui, "Transhumanism: Meet the Cyborgs and Biohackers Redefining Beauty," CNN, May 27, 2020. https://www.cnn.com/style/article/david-vintiner-transhumanism/index.html.

Index

A
Adib-Moghaddam, Arshin, 132–134
aerospace industry, robots in, 77
Ahmad, Imran, 34, 36, 37, 39
Albright, Jonathan, 146
Alonso, Cristian, 29–32
Ansari, Atif, 36–37, 38
artificial intelligence (AI), definition of, 13, 52
Asilomar AI Principles, 135
Asimo, 78–79
autocorrect/text editors, 52

B
Bartlett, Albert Allen, 107
Baum, Sandy, 98
Beardslee, Susan, 161
Berlin, Isaiah, 59
Bernier, Chantal, 33, 35, 36, 38
Blankendaal, R. A. M., 62–75
Boer-Visschedijk, G. C. van de, 62–75
Bonilla, Carlos, 116–120
Boonekamp, R. C., 62–75
Born, Kelly, 150
Brandeis, Louis, 151–152
Brynjolfsson, Erik, 107
Burke, Marshall, 42–43

C
Carnevale, Anthony, 97
chatbots, 54–55
Ciarniello, Alex, 154–157
Clark, Dave, 123–124
cybersecurity, AI and, 135–142

D
deep learning, 19, 53, 54, 55, 85–86, 129
DeepMind, 18, 21, 85
Dickson, Ben, 85–89
digital/virtual assistants, 55, 112

E
education, and AI, 44–51, 95–102, 162–167
Eikelboom, A. R., 62–75
entertainment industry, robots in, 78
e-payments, 56
Etzioni, Oren, 126–131

F
Facebook, 66, 133, 144, 145, 148, 150, 151
facial detection and recognition, 54
food production/poverty, and AI, 40–43, 105–106

G
Giannini, Stefania, 163
Goddard, William, 44–51

H
Hawking, Stephen, 14, 90–93
Holzer, Harry, 98
Horvitz, Eric, 27–28
household appliances, robotic, 77–78

I
iCub, 79–80
independent robotic security systems, 77

J
jobs, effect of AI on, 14–15, 20–21, 29–32, 95–102, 103–111, 112–115, 116–120, 121–124

K
Kantor, George A., 41, 42
Kaste, Martin, 23–28
Korteling, J. E., 62–75
Kothari, Siddharth, 29–32
Kuratas, 79
Kurzweil, Raymond, 59

L
Lazer, David, 146
LeCun, Yann, 66
Levesque, Elizabeth Mann, 95–102
Lokitz, Justin, 121–124

M
machine learning (ML), 17, 27, 29, 33, 40, 41, 42, 44, 45, 46, 48, 50, 51, 53, 54, 55, 85, 98, 121, 133, 134, 135, 136, 137–138, 140, 154, 155, 156
Maddox, John, 159, 160
Maddox, Teena, 158–161
manufacturing, robots in, 77–78
maps and navigation, 53
Mason, Elisabeth, 40, 43
McClelland, Calum, 103–111
medicine, robots in, 76, 92–93
Milford, Michael, 17–22
misinformation/fake news, using AI to combat, 143–153
Moravec, Hans, 59
More, Max, 58, 59
Morton, Carrie, 160
Murray, Jasen, 28
Musk, Elon, 14, 17, 93, 126, 127–128

N
national security, AI and, 154–157
natural language processing (NLP), 53, 54, 55
Nyhan, Brendan, 148

O
Ovaska, Sarah, 33–39

P

Pennycook, Gordon, 151
Powell, Wayne, 159
privacy risks of AI, 33–39
Progressive Automations, Inc., 76–84
Putin, Vladimir, 126

R

Rand, David, 151
Reeves, Sasha, 52–56
regulation of AI, 15, 126–131
Rehman, Sidra, 29–32
robot companions, 81–82
Roboy, 79
Roepe, Lisa Rabasca, 40–43
Rollin' Justin, 80
Rutschman, Ana Santos, 90–93

S

search and recommendation algorithms, 54
self-driving/driverless cars, 18, 19, 87, 114, 118, 127, 130, 158–161
Singularity, 23–28
social media, and AI, 55, 127, 143–153
Spaeth, Dennis, 112–115
Stith, Paul, 161
Stratton, Peter, 17–22

T

Tacchini, Eugenio, 150
Thiel, Peter, 26

transhumanism, 15, 58–61
Trippett, David, 58–61
Twitter, 128, 144, 145, 146

U

UNESCO, 162–167

V

VGo, 79
voice/speech recognition, 17, 18, 19, 52, 86, 118

W

Wang, William Yang, 150
West, Darrell M., 96, 143–153
Wheeler, Tom, 149–150
Wilkins, Benjamin, 122
Wolff, Josephine, 135–142

Y

Yudkowsky, Eliezer, 26

Z

Zannoni, Mark, 158, 161
Zimdars, Melissa, 148